G.I. Hollywood

Some of My (Mis)Adventures in Show Business and the Military

By George Mannix
With Lila McLaughlin

My journey from Hollywood Agent to Combat Soldier,
and my return home into Love and Marriage...

Table of Contents

Disclaimer

The views expressed in this memoir reflect those of the author and do not reflect those of the Army National Guard, the United States Army, the Department of Defense, or the United States Government.

For Dawn and Meghan

Notes from the Authors

"Life moves pretty fast, if you don't stop to look around once in a while, you could miss it." – Ferris Bueller, <u>Ferris Bueller's Day Off</u>

This book is not the story of my life. This book tells some of the stories that are *part* of my life. My work life consists mostly of experiences in Hollywood and the Army. This book discusses some of my best and worst times in both of those worlds. In addition to my work endeavors, this book also discusses how my wife and I met, got married, and had a daughter – the joy of our life.

I was a Plebe at West Point, completed 1L at Law School, worked as an agent in Hollywood, and I served in combat in Baghdad, Iraq. Along the way, I met many different and very interesting people, from both show business and the government. Some of these people encouraged me over the years to tell some of these stories.

The stories included in this memoir are about my perspective of the events described and are not intended to represent anyone else's views or memories.

Through all the trials and tribulations contained herein, I generally looked forward rather than look back. I firmly believe that my sense of optimism enabled me to enjoy the life that I have lived thus far.

Thank you for allowing me to share these stories with you. I hope that you enjoy them and learn from them. To be continued…

All my best,

George Mannix

In 2003, George Mannix had to leave his career as a young Hollywood agent to go to Iraq and fight in a war. Yet, his journey was much more than that. There are many layers to his fascinating story. Mainly, one of finding purpose amidst the chaos life often serves up.

I wanted to write George's story because it relates to anyone who has ever been faced with a big decision, or a challenge, where they have to leave their comfort zone because they made a promise, or not had much of a choice. But then, that which they were reluctant to go and do, brings that person much closer to being happy and finding some purpose.

Ultimately, this incredible story is not rooted in just show business or the army. But rather, it's humor, charm and heart convey an overall universal theme that everyone has to find their own path to happiness.

His encounters and experiences are told in a journal format to give a more active account of his journey.

This is not a book about politics, or opinions on the war, and we (the authors) have no political agenda or sway in telling his story.

It is solely a memoir George and I put together of his accounts in two very different worlds. His journey from dealing with studios and making film deals as a Hollywood agent, to Soldier for the

U.S. Army, where he was commanded to do a duty to protect and serve for this great nation.

Take care and enjoy,

Lila McLaughlin

Prologue: The Negotiation (that Never Happened)

"There are three sides to every story. Yours. Mine. And the truth.
And no one is lying. Memories shared serve each differently." –
Robert Evans, <u>*The Kid Stays in the Picture*</u>

August 2002

"Negotiate with them!" The Boss told me.

It was August 2002. I was working at the The Agency, a Talent
and Literary Agency in Los Angeles, California. The Boss was the
chairman and owner. To people that knew him, which included
nearly the entire Hollywood galaxy, his friends, his family, and the
people that worked for him, he was an agent from the "old school"
of agenting. I told The Boss that I was out of law school and
reporting for active duty in the Army.

A little back-story…

On May 20, 2001, I was commissioned as an officer in the U.S.
Army through the Reserve Officers Training Corps at the
University of Rhode Island. I was accepted to law school at
Southwestern University School of Law in Los Angeles,
California. Due to this, the Army granted me an "educational
delay" from active duty. In exchange for the educational delay, I
was obligated to serve three years on active duty upon completing
law school, presumably as a member of the Judge Advocate

Generals (JAG) Corps – the military lawyers. I was commissioned in the Infantry.

Now, back to Hollywood…

"You're an executive. You belong behind a desk," he said.

"I appreciate that, but this is the United States Army. Negotiating with them is not an option. I have orders," I said.

"You know, I was in the Navy Reserves," he told me.

This was something he never told me before. He continued, "After I completed my time, they asked me to come back in so that I could go to the Korean War. You know what? It didn't happen."

"Understood," I replied, "but I committed to three years on active duty *before* I came to Los Angeles for law school. I owe Uncle Sam."

"I see," he told me. "Let's get back to work."

It was tough to have to leave a job that I had worked nearly three years to get. It was tough to leave The Agency. It was tough to leave The Boss. It was even tougher to leave show business. But I had orders and my orders directed me to report to the United States Army.

I soon learned that the Army would "re-assess" me. This meant that I might be assigned outside of the Infantry. The simple solution would have been for the Army to issue me orders sending me to Fort Benning, Georgia – the Home of the Infantry – for training in the fall. Instead, it would take nearly six months.

In the meantime, I had to wind down my affairs in Los Angeles. In time, I would depart The Agency. I would terminate all my affairs in the film and television industry, and I would prepare to be a Soldier.

Chapter 1: How It All Started

"We Didn't Start the Fire / It was always burnin' / since the world's been turnin' / We didn't start the fire / No we didn't light it / But we tried to fight it" – Billy Joel, "We Didn't Start the Fire"

February 1979 – September 1998

Growing up, I was a cocky son-of-a-bitch. From the beginning, I was willing to duke it out physically or verbally with anyone that I thought was unfair or a bully in some way; this included authority figures such as teachers. As a kid, my mom used to have a pep talk before the first day of school. It went something like this, "Say 'please' and 'thank you' – remember your manners. If you see someone that needs help, then give them a hand. However, if some kid pushes you around or hits you, HIT THEM BACK! I will back you up all the way."

I essentially followed that to a T.

My dad would say, "Be a good kid and be a good student. I wish that I had done better in school as a kid. And…do what your mom told you…"

Dad was pretty easy-going, very fair-minded, but strong. He inspired me to do my best.

My father, Michael, was a quality control manager. He had worked his way into management after being a welder and supervisor on the floor at Electric Boat. The General Dynamics Electric Boat Division was responsible for the construction of nuclear submarines fo the U.S. Navy. The purpose of the submarines was to enable the nation to be ready to strike the Soviet Union from the sea quickly and without detection, if need be. The submarine was a critical element to the arsenal of the Cold War. My father was also a veteran. He was in the U.S. Navy from 1965-1969.

My mother, Patricia, was a professor of English at the Community College of Rhode Island. She worked there for forty years. My brothers, there were four us, and I grew up in Narragansett, Rhode Island. Narragansett sits in the bay sharing the same name. Most mornings, I could see the sunrise over the ocean. Narragansett is a town of ten thousand from Labor Day to Memorial Day and about thirty thousand people during the summer. Tourism and fishing were the biggest industries in town.

My hometown (Narragansett, Rhode Island) is also well within the confines of "Red Sox Nation." As a kid, we were all pulling for the Red Sox to break "The Curse of the Great Bambino" – the sale of Babe Ruth to the New York Yankees after the 1918 World Series.

My oldest brother is Patrick. He is proud to be an Irish-American and loyal. He aimed to help all my brothers and I do well in life.

Matt, is my second eldest brother. He is organized, smart, and out-going. He enjoys bringing people together and solving problems.

Chris is my little brother. He is extremely intelligent, energetic, and well-traveled. He loves being with people and getting right into the heart of anything that is going on.

For the most part, my brothers and I got along well growing up.

Due to a solid foundation at home and a good work ethic, I did very well in school. I also did well in sports and other extracurricular endeavors. From early on, or at least since I could talk, I was also good at taking charge and leading people. This included everything from the games on the playground to organizing the lunch line.

So when it came time to figure out life after high school, I realized early into my senior year that one of my options was to attend the United States Military Academy at West Point, in New York. From an early age, I was interested in serving my country, so attending college at West Point was a strong consideration for me. Senator Jack Reed from Rhode Island personally called me to advise that he appointed me to the Academy. He was himself an "Old Grad" –

a member of the Long Gray Line – West Point Class of 1971. This was truly an honor. It was also a surprise.

My love for the military came from my earliest days of childhood. G.I. Joe was my favorite toy. Being in the Army also seemed like a very action-filled lifestyle. Beyond that, my family instilled me with a love of country. As I said, my dad served in the U.S. Navy before I was born and my mom always had the flag out front. Our home was a "God, Country, and Family" place.

My other great passion was entertaining. I was very good at impersonations in my teen years. I used to impersonate movie stars and singers for my friends. I think this artistic side was influenced in part by both of my parents. My dad was very creative with his hands. He could make just about anything in wood and spent much of his free time doing woodworking. My mom was an extremely talented pianist. She still can play just about any song by sight.

I later directed my class's variety show during my junior year of high school and in my senior year, I co-wrote, directed, and acted in a play called *Good Mariners*, which blended *GoodFellas*, *Casino*, and local jokes. For both the variety show and *Good Mariners* to happen, I butted heads with the adult school leadership. At the end of the day, my feeling was, if the students want this and we can execute it without interfering with the big plans of the "grown-ups," then let's get it done. Needless to say, my "take it or leave it" approach to getting things done did not go over well with all parties involved, but I didn't care. We had a show to produce.

These experiences also allowed me to work with all kinds of students at the school. The "grown-ups" also viewed some of the students that supported these activities as troublemakers, but they did incredible work for myself and the show. This reinforced, in my mind, the importance of both fairness and giving everyone a chance (or second chance). I was demanding, stubborn, but also very loyal to my cast and crew and I believe that I was also very fair.

I loved seeing people laugh. The play ended up being a huge success. The audience started standing after the second wave of the cast started taking their bows. By the time all the leads bowed during the curtain call, the entire audience was on its feet. We (the cast and crew) enjoyed a lot of support from the entire student body. Two touching things happened:

1. A bunch of students (mainly girls) came in 2-3 hours before the show started in order to decorate the auditorium. I hadn't asked them to do this and was very touched, they did a wonderful job.

2. Unbeknownst to me, the father of one of my classmates taped the entire play (he had once been a little league coach of mine too). This was also a nice surprise.

From the show being a success and all the buzz it created, both the school newspaper and the town newspaper interviewed me about the play. In my interview with the town paper, I was asked whether I would one day go into show business. At the time of the interview, it was almost definite that I would be heading to West Point for college. I told the paper that I'd most likely work in the military or the government.

I was fortunate to have received, relatively early admission to the United States Military Academy at West Point, New York. Because of this, I didn't apply to a great number of colleges and universities.

Two weeks after my high school graduation, I went from, "High School hero to West Point zero." Until you complete the first summer of training ("Plebe Summer") you are referred to as a "New Cadet." At the completion of "Plebe Summer," you are referred to as a "Plebe." Plebe equates to a college freshman. "Yearling" or "Yuk" is the term for a sophomore. "Cow" is the term for a junior. "Firstie" is the term for a senior. All upperclassmen are called "sir" or "ma'am" (as if they were superior officers in the Army) by Plebes until "recognition" which occurs in the middle of the second semester. Following recognition, all cadets are on a first name basis. The actual military officers assigned to oversee the cadets are called tactical officers. They are addressed as "sir" or "ma'am," consistent with military courtesies. Many faculty are also military officers. All faculty were addressed as "sir" or "ma'am."

As a Plebe, you are required to be in uniform 24/7 while at or around West Point. There is literally a uniform robe and shower shoes for bathing, along with government issued towels and washcloths. There is a physical training uniform, which doubles as pajamas. There are several uniforms. Most of them are gray and black, which are worn for class, training, parades, etc. Humility is one of the first lessons you learn and it is provided "the hard way." At West Point, you actually are paid a small stipend to go to school from the beginning. As the cadet saying goes, "You get paid up the ass, one nickel at a time."

At West Point, I continued to engage in theatrics often. This had the effect of either amusing or pissing off my superiors. Depending on their humor, it either meant that I was "the funny guy" or "the grand jackass." One upperclassman thought that I sounded like Jack Nicholson, so that upperclassman would come up to me, and say, "Talk!"

Then, he would tell me something like, "My God, you sound like Jack Nicholson."

Other upperclassmen, however, would say, "Why are you making jokes? The Army is a serious profession. Don't be a clown!"

To them, I simply replied, "Yes, sir."

At the end of Plebe Summer, there was a talent show. In the summer talent show, I performed in two skits. In one of them, some of the other cadets and I did a sketch of the scene between Tom Cruise and Jack Nicholson from *A Few Good Men*. I played the Jack Nicholson part – Colonel Jessup. The other sketch that I did was an impersonation of General George S. Patton, Jr. from the beginning of the move *Patton* – the famous speech before the 3rd Army. The Superintendent allowed me to borrow his rank (his three stars) in order to do the impersonation. The Patton speech went over so well that I ended up performing the speech again in December before the annual Army-Navy football game.

The Patton speech was recorded and later it was used in videos for incoming cadets.

Over the course of my Plebe year, one of my friends, I later referred to as LA Marvelous, became the person that all of the upperclassmen picked on. LA Marvelous got his name from doing stand-up comedy (after he left West Point). He was tenacious, funny, and very common sensical. I believe that LA Marvelous' desire to see the logic in all things was part of what sparked the ire of the upper classmen.

In addition, the upperclassmen punished him more severely than the rest of us. When we were chewed out, LA Marvelous would be formally reprimanded. This was constant. I viewed this as incredibly unfair. His infractions gave rise to a disciplinary hearing. By the rules governing the hearing, he was allowed to have legal counsel and ask questions of the hearing officer. He was denied both. I testified as a character witness on his behalf. I went in, spoke my piece, and started to lay down some of the incongruities with how he was treated. The hearing officer summarily dismissed me. When LA Marvelous went in, the hearing officer did not allow him to ask questions. I told the officers supervising us that the execution of the proceeding was all wrong. They didn't seem to care. I lost heart in the "Long Gray Line" (being a West Pointer) then and there. A couple months later, I left West Point.

In time, I came to have a renewed respect for West Point. Of course, I still have many friends that graduated from the academy. I also believe that it plays a very important role for our Army and Nation.

But ultimately, it just wasn't for me.

The downside of leaving West Point shortly before Labor Day of my Yearling (sophomore) year was that my next big gig was.... Drum roll... college dropout, at least for a semester.

I was out of school, out of the military, and out of a job. The upside was that I had a clean slate.

I transferred to the University of Rhode Island and resumed college in the second semester. I majored in history, with a focus on American history during the 1920s.

I also enrolled in the Reserve Officers Training Corps (ROTC) program there – the Cramer's Sabers Battalion, headquartered at the University of Rhode Island. ROTC with the Cramer's Sabers played an enormous role in my life.

The namesake of Cramer's Sabers Battalion was 1st Lieutenant Parker Dresser Cramer. 1st Lieutenant Cramer was killed in action in Vietnam and the University of Rhode Island named its ROTC Battalion in his honor.

In spite of leaving West Point, I still wanted to serve my country. Cramer's Sabers Battalion gave me a framework for moving forward. I served as one of the leaders in the battalion and I was extremely proud of what we accomplished as a battalion when I was a cadet. My understanding is that we went from the bottom half of ROTC programs in the country to the top ten percent in two years. Obviously, such progress was a team effort, but I was proud to have been one of the cadet leaders during this time. I also enjoyed the company of my fellow cadets and some of them are still close buddies. Cramer's Sabers Battalion renewed my faith in the Army. Through ROTC, I was once again re-integrating with one of my interests, the Army.

But by being in a civilian university rather than a military academy, I was also able to explore other passions. So, I took the opportunity to re-connect with my other interest, film studies and the business of movie-making.

I dropped an application to attend the University of Southern California's School of Cinema-Television (now the School of Cinematic Arts) for a Summer Production Workshop. I was accepted. I figured that summer film school would tell me if I liked Los Angeles, since I had never been there before. I also figured that it would tell me if I would like filmmaking and the movie industry.

Chapter 2:   The Magic of the Movies

"Louie, I think this is the beginning of a beautiful friendship." –
Rick Blaine, Casablanca

More than anything, what sparked my passion for filmmaking was
experiencing the magic of the movies with my family, a bucket of
popcorn, a cup of soda, and images flickering across the screen in
surround sound at the cinema. Although, we didn't go to the
cinema often, my family and I always had pizza and a movie on
Friday nights while growing up. When we did go to the movies, I
loved it. Five movies greatly influenced me – *E.T.*, *Return of the
Jedi*, *Superman*, *Casablanca*, and *Black Hawk Down*.

E.T. the Extra-Terrestrial

1982

E.T. is the first movie that I remember seeing on the big screen. As
Elliott and his siblings try to figure out how to both hide and
nurture E.T., I was amazed. To some degree, they were playing
"hide and go seek" with their mother and then later with the
government. The idea that a little boy could help this special alien
touched me. Near the end when they get to the woods and fly their
bicycles across the moon, it was magic. Not only did the kids win,
but they also got to fly!

Return of the Jedi

1983

Return of the Jedi is another movie that I vividly remember watching at the cinema. The opening titles to the *Star Wars* film backed by John Williams' timeless theme captures you instantly. What *Return of the Jedi* (and all of the *Star Wars* movies) portrayed is the classic struggle between Good and Evil. The good guys – the rebels – put together a plan, built a team, and fought their enemy. The good guys won. But *Return of the Jedi* was not only about Good triumphing over Evil, it also demonstrates that certain things are worth fighting for like freedom and your way of life.

Superman

1978

Superman was released on the big screen before I was born. When I was a little boy, it was available on VHS. In the late summer, when I as five years-old, I was hit by a car. I remember waking up in the hospital to *Superman.* Superman was and still is my favorite super-hero. He stands for "Peace, Justice, and the American way" similar to GI Joe being "The All-American Hero." My leg was broken and I had suffered a concussion. Seeing Superman fly around and save Metropolis made me feel better and want to get well as soon as possible. Superman used his powers to protect people from danger and fight evil. Superman made it easier for Americans (and the world) to enjoy life by being present in our world. Superman was a force for good, no matter how bad the evil he faced. And Superman won.

Casablanca

1942

Casablanca is my favorite movie. I originally saw *Casablanca* on VHS as a boy in the 1980s. When I turned twenty-one, *Casablanca* was screening at the Providence Performing Arts Center (in Providence, Rhode Island). I went to see it on the big screen.

For the audience (past, present, and future), *Casablanca* has a great deal to offer. It has the quintessential romantic triangle. The setting is exotic. There is good versus evil against a very real backdrop (World War II). The characters face difficult choices. Many characters must choose between what is right and what they themselves want. The American – Rick Blaine – is independent, individualistic, and ultimately the most reluctant of heroes.

Technically, *Casablanca* is a masterpiece. The story is compelling and both major and minor characters are flawlessly brought to life. The cinematography and the production design are beautiful. The editing optimizes every word and image. The music amplifies both the story and the characters. The film's themes are timeless.

For me, *Casablanca* is wonderful because it shows how interconnected we are and how important we can be to one another and even to the world. Rick Blaine, the film's protagonist, is also the archetypical American. Rick is accepting of others. Rick aims to be a successful businessman with his café. Rick doesn't want to be engaged in the affairs of the world, but he also doesn't want to see people suffer. Rick is loyal, generous, and righteous. This and other elements are what I love about *Casablanca*, including the fact that it is set against one of mankind's greatest battles between good (the United States and our European Allies) and evil (Nazi Germany). The outcome of World War II was unknown when the movie came out. Fortunately, for our country, the world, and us, the Allies prevailed due largely to American leadership in the war. The strength of one man, an American hero, and the American "way" are all central to the movie, and something I enjoyed.

Black Hawk Down

2001

Black Hawk Down came out in 2001. The 9/11 attacks had just happened and our country was now at war. While the movie is set in Mogadishu, Somalia in 1993, the story of a small-unit in a close, urban battle was very relevant. I was a recently commissioned

Army officer and I was midway through my first year in law school when I saw the film. *Black Hawk Down* represented what my future could hold. The film brought to life in a very graphic manner the realities of war to Soldiers. Today, I still consider *Black Hawk Down* to be the most matter-of-fact portrayal of modern combat that I have ever seen. The film also captures the tight-knit nature of a small unit thrust into battle. It captures both the fog of war and the importance of leadership in lifting at least some of the fog. On a grand scale, it portrays the most solemn decision point that our civil leaders face, which is understanding the consequences, the risks, and the goals of going to war.

$$*****$$

E.T., *Return of the Jedi*, *Superman*, *Casablanca*, and *Black Hawk Down* touched me in different ways. They all helped shape one of my life goals, to work in the movie industry as either an agent or a creative executive at a studio or production company.

Agents represent the talent that make the movies and in show business function as "the sellers." Creative executives develop and produce movies and in show business function as "the buyers."

Before my big screen viewing of *Casablanca*, the magic of the movies touched and entertained me as they would most people. But by the time I got to experience *Casablanca* on the big screen, that magic drove me to want to be part of the process of making movies.

That drive would lead me to Los Angeles, California.

Chapter 3: Creating My Path

"Got on board a westbound Seven-Thirty-Seven / Didn't think before deciding what to do / All that talk of opportunities / TV shows and movies / Rang true, it sure rang true." - Albert Hammond, "It Never Rains in Southern California"

June 1999 – May 2002

Film School

June – July 1999

In the summer of 1999, I attended the University of Southern California / Universal Studios Producing and Directing Summer Seminar. It was a five-week program that was designed to teach students how to write, shoot, and edit films, as well as introduce the basics of show business. Part of the introduction to the basics of show business entailed going to Universal Studios every Wednesday in order to "meet the industry."

Our class met producers, directors, writers, agents, cinematographers, editors, and production designers with major credits to their names. All of these industry folks were working on projects and they were there to share their industry "war stories" with us.

One of the class's most frequent questions was, "What does it take to succeed in the movies?"

One producer said, "To be honest, you have to be a whore."

He went on to explain that what he really meant was that you have to get along with people, especially the studio brass, the investors, and the agencies that are involved with packaging your film projects.

A director replied, "You need to know people. And you need determination."

A writer said, "Networking and persistence."

At the end of the class, it was clear from the industry veterans that key to establishing and maintaining a career in show business were persistence and resilience.

During the days when we didn't visit the studio, we learned the basics of writing, shooting, and editing film. We each had to direct two projects. Most of the time was spent working on other people's projects. For me, that meant acting.

In addition to the lessons about persistence and resilience that were learned, the on campus portion of film school taught me that I was not a natural-born director. I understood story, I understood characters, and I understood putting the key pieces together, but I did not have "the eye" that so many film directors have.

What did this mean? It meant that some day I would become either a producer or an agent. In those jobs, a good sense of story, character, people, and business savvy are critical to success.

But aside from having the natural skill to be a good producer or agent, I also did well on screen at film school. Two of the student films that I acted in were selected to go to the festival that featured the "best student made films of the workshop" at the end of the seminar. This brought together the best two movies from every class that was conducted during the summer seminar.

Only two films were selected from each workshop group. Since the student demographics ranged from highly talented high school seniors to working professionals in show business, there were many different eyes on the movies at that festival. I acted in both of the films that were selected for the festival from my class, so I got the film school "star" treatment at the festival. It was a blast and remains my favorite five weeks of college.

In spite of the success of my "acting" in the student films, I still wanted to be either a producer or agent. While acting was fun, it also brings with it the potential of celebrity – something that was of no interest to me. Rather, I truly enjoyed building stories and teams, behind the scenes, bringing stories to life.

More importantly, during my brief time at film school, I learned that I loved (not liked): Los Angeles, filmmaking, and the entertainment industry.

The clincher was the sunset at the beach in Marina Del Rey – it was amazing and warm. Growing up on the East Coast, I literally watched the sunrise over the ocean many times. However, out West, seeing the sunset over the ocean was incredible. I think in some ways, it gave me the feeling of the "movie hero" riding off into the sunset.

During that trip, I knew that once film school was over, I wanted to get back to L.A. and set conditions that would allow me to have a life and career in show business. I started formulating a plan to learn more about the movies. Then I put together a timeline on when I'd return to Los Angeles after college and the Army… to see many more sunsets in Marina Del Rey.

Let's Talk

July 1999

After I completed film school, it was time to head back to Rhode Island. I got on the plane at LAX and got ready for a long flight. I was quite happy. I felt that my time at film school had been informative and exciting.

"Any one sitting here?" The man next to me asked.

"Nobody. Be my guest," I replied.

He took the seat next to me. He was middle-aged, with reddish hair, and a short beard.

"Pretty nice day," he said.

"Well, it seems that it is always seventy-five, sunny, clear, and dry out here," I said.

"Were you here long?" He asked.

"Five weeks," I said. "How about you?"

He just seemed to skip over my question. "So, what brought you out here?"

With a small degree of hesitation, I told him, "Film school."

"Really? Where?"

"The University of Southern California."

"That's impressive."

"Thank you. It was fun."

"Well," he begins to tell me, "I think that some of the work that I do is sort of like writing movies."

'No shit,' I thought to myself, 'that's what everyone says in La-La Land. I should've listened to what the folks at Universal Studios told us – there are a lot of people that want in.'

Politely, I replied, "How so?"

This will be a long flight, I thought.

"I write for a newspaper. We have to figure out what is interesting, who the players are, that kind of thing. It's like coming up with a story, except you tell people about things that really happened."

"So, what kind of stories do you write?"

"All different kinds. One of the recent stories that I wrote was about these folks in India…"

As he told me the story, I realized that he was not writing for a small-town paper.

"What paper are you with?" I asked.

"The Baltimore Sun," he said.

"That's impressive," I told him.

He was an ace investigative reporter for *The Baltimore Sun*. We talked for the entire flight. Shortly before the plane reached Baltimore, his stop, he told me that the story that started with 'these folks in India' recently won the Pulitzer Prize.

"That's a selling point," I said. "You need to make sure people know that, if you want to make it into a movie."

"I want to (make it into a movie)," he said, "but it is hard to figure out who to talk to and how to move it forward and all that stuff."

"I tell you what. Let me check with some of the people that I met over my time at film school and see if we can figure something out for you."

"Sounds good," he said. He gave me his business card. We then introduced ourselves and shook hands.

"Safe trip," I told him, as he left.

An hour later, my plane was in Providence and I was back in Rhode Island. My odyssey to Los Angeles was complete for the time being.

A few days later, I called the Pulitzer Prize winning reporter that I had met on the plane.

"Okay, here's the deal. You know that I just got back from film school and that I have a couple of bona fide contacts in the industry. If you are open to it, then what do you think about this – I represent you and the story and we see what happens?"

"Sounds good to me," he said.

I was shocked.

I was twenty years old. I had just started out as a free-lance rep for film and television talent. The Pulitzer Prize winner became my first client.

Now, my job was to start making things happen.

Airborne

July – August 1999

For a couple weeks after returning from film school, I was working prospects for "Pulitzer" (as I had dubbed him). I made some calls to producers, studios, just about anyone in the industry that would listen to me. Then I made some more calls and some more calls. I followed up every call with e-mail. So far, so good – everyone seemed "interested in learning more about the project."

Then, the senior non-commissioned officer from my ROTC Battalion called.

"What's going on?" I asked.

"There's an opening at Jump School. Do you want to go?" He said.

"Sure, when do you need a final answer?"

"Today, we got this on short notice."

In ROTC, the senior non-commissioned officer is responsible for ensuring that everything within the battalion gets done. So, when a school date opened up, his job was to ensure that it was filled or passed on to another battalion.

"Sergeant, I will go," I told him.

"Sounds good," he said. "Out here."

A day later, I get a call from Angie, a friend I had met while at film school. She was now working in casting. Angie was energetic and athletic. At 20, she was our film class's "veteran" in show business. She worked as a production assistant on several film and television projects.

Angie and I worked on several short films together at summer film school. She was crew on the first student film that I made and part of the cast on the second. I crewed both of her films. We acted together in several of our classmates' films. The most memorable was a science fiction movie called "Project: Gemini" where she is cloned and her evil clone tries to kill her, but I kill the evil clone and save her. We became friends and enjoyed working together. For whatever reason, we clicked.

Angie told me that she was working with Tony and Ridley Scott and also on some television projects.

"That's great," I told her.

"Here's the thing," she said, "I showed some people your reel."

"Okay."

"And they liked it," she said.

"Okay." Go on, I was thinking,

"And they want to know whether you want to do cigarette commercials in Europe. This will probably be with Tony and Ridley Scott," she said all in one breath.

I was about to say, "That's great."

"But," she continued, "I am also working on a pilot episode for a new TV show."

"Okay," I said, still interested.

"The show will be called "The West Wing" and they are interested in you as a possible extra. The idea here is that you get in on the pilot and if the series gets picked up, then maybe you can have a part on the show," she said.

"This is unbelievable. You are awesome," I said.

"So, are you in?" She asked.

"Hell, yeah, when does this start?"

"Next week?" She said.

"Oh," I said, "and when does this end?"

"By the end of the summer."

"That's not so good," I countered.

"Why?"

"Well, I just promised the Army that I would go to Airborne School during the next three weeks."

"So, you will jump out of planes?" She asked.

"Yes," I said.

"That's cool."

"I guess so. Do you think maybe something will come up later?" I asked, hopeful.

"It's always possible, Kid," she said. She called everyone "Kid."

"Take it easy, Kid."

"You too," I said.

I hung the phone up. Very few times in my life have I regretted being decisive. Initially, this was one of them.

By seizing the opportunity to go to the U.S. Army Airborne School, I was declining the opportunity to screen test for television, and possibly the movies. While being on screen was not my ultimate goal, it certainly was one way to get into show business.

Nevertheless, I made a promise to the Army to go to Airborne School. That took me to Fort Benning, Georgia for the first time in my life.

Three and a half weeks later at Fort Benning, Georgia, my dad pinned my silver wings on after I completed my fifth jump at Airborne School. I had gotten pretty sick while I was there, but after turning down the acting opportunities, I was bound and determined to get those wings. I was proud and so was Dad. In fact, my medic from Plebe Year at West Point was there too. He came over to congratulate me. Maybe I wasn't on the path to television stardom yet, but I was doing pretty well on the road to becoming an Army officer.

Upon returning to the University of Rhode Island that fall, I was a junior. Due to that, I had to decide whether to commit to the Army. I committed to the Army.

At this point, I was ready and willing to make this commitment. After going to and leaving West Point and then joining ROTC, I knew that I still wanted to serve my country. My commitment was to serve the Army as an officer.

If I did not successfully complete ROTC, then I would be required to serve three years on active duty in the Army as an enlisted Soldier. This obligation was laid out in my contract.

The Army *and* Show Business – My Way Ahead

September 1999 – May 2001

My plan evolved, to pursue both my interests. While I was completing college and ROTC, I continued independently representing Pulitzer and I added a Tony nominee and two Emmy winners as clients too.

I wanted to be commissioned in the Infantry and detailed to the Judge Advocate Generals (JAG) Corps. I figured that this would allow me to do the following:

(1) The detail to JAG would allow me to go to law school (if I was accepted).

(2) I wanted to go to law school in Los Angeles, to stay close to the entertainment industry, and get a toehold in "the Establishment."

(3) Then, I would serve my three years in the Army as a lawyer.

(4) After that, I would return to Hollywood and build upon my experiences from my law school tenure, and ultimately, secure a position in the entertainment industry.

Cramer's Sabers' Class of 2001

May 2001

On a warm and bright spring day, I joined ten of my fellow brothers and sisters of Cramer's Sabers Battalion and was commissioned. I was an Infantry officer in the United States Army. In doing so, I took the following oath:

> "I, Michael George Mannix, having been appointed an officer in the Army of the United States, do solemnly swear that I will support and defend the Constitution of the United States against all enemies, foreign and domestic. That I will bear true faith and allegiance to the same; that I take this obligation freely, without any mental reservation or purpose of evasion; and that I will well and faithfully discharge the duties of the office upon which I am about to enter; so help me God."

I was selected by the Cadre to speak to my fellow classmates. I talked about the gravity of the duties that we had just sworn to uphold. I also talked about the beauty and the bounty of the United States of America. In addition, I talked a little bit about the time we spent together as cadets.

We were the final class to become officers before a prolonged period of combat operations. At the time, the most recent conflict was in the Balkans. War seemed very abstract to us, but certainly possible. Within four months, our Homeland was attacked. War would be upon us all.

My parents pinned the gold bars of a 2nd Lieutenant onto my uniform. I received my first salute from my father. We could do this, because he had once served in the U.S. Navy as a non-commissioned officer. I was proud and happy. Later, that very same day, I graduated from the University of Rhode Island.

L.A. Law

August 2001

Three months after I was commissioned an Army officer and graduated from college, I was on a westbound plane. I flew to Los Angeles, California once again. This time, I was en route to start law school.

I had continued representing writers on a free-lance basis. I made contact with producers and studios, and I also worked with the writers on improving their treatments, scripts, and books.

By going to law school, I was still on track with my plan. I was commissioned in the Army as an Infantry officer and was detailed to JAG, the path to being an Army lawyer. Due to the JAG detail, the Army approved of me getting an educational delay. I was given three years to get a law degree. Then, I had to report to active duty. Contingent upon successfully completing law school and passing the bar, I was committed to serve as a JAG in the Army.

I also decided that I wanted to go to law school in order to better position myself for Hollywood after my military duty was finished.

My goal was two-fold in going to law school. I figured that it would provide me with the opportunity to get my foot in the door of Hollywood and that holding a law degree would set higher conditions for employment at an agency or studio later. I also figured that it would be helpful to learn about business organizations, contracts, intellectual property, and entertainment law, if I was going to work in "Show Business."

I had previously applied to five law schools in L.A. that were accredited by the American Bar Association. I was was accepted by one – Southwestern University School of Law.

My plan was going along swimmingly; I was representing talent, building my client roster, and I was about to start law school in the city where I wanted to be.

9/11

September 2001

Two weeks into my law education, also known as "1L," I was watching the news in the early morning. A plane had just flown into one of the Twin Towers of the World Trade Center in New York. As reporters were searching for answers as to why that happened, another plane flew into the other Twin Tower. It was very clear that our Nation was under attack.

The last thing that I wanted to do was go to class. I felt like calling the Army and reporting for duty. I reluctantly went to class. Several students were trying to reach family and friends on the East Coast. There were many difficulties with communications getting through to New York City and Washington, DC. I actually had two brothers in Washington, DC at the time and the Pentagon had also been hit. I heard that news on the radio en route to school.

I recall being impressed with the fact that Los Angeles quickly deployed both helicopter patrols and police ground patrols in high-traffic areas very quickly (probably within an hour or so of the first plane hitting the World Trade Center). I figured that earthquakes, wildfires, mudslides, and the not-so-distant riots had prepared the city well for executing contingency operations. It took a couple of days before I received confirmation that my brothers in

Washington, DC were safe and sound. My brothers got to a safe place – their friends' house in Falls Church, VA. After a few days, cell phone service was restored. I heard by way of my parents that my brothers were fine.

It took a while for family, friends, and the Nation to get a clear idea of what happened.

As a teenager, I traveled to New Jersey and Washington, DC many times from Rhode Island to visit my brothers at college and later their new homes. Traveling south on the New Jersey Turnpike, the Manhattan skyline comes into view to the east. The most distinguishing feature was always the Twin Towers of the World Trade Center. It was hard to believe that the Twin Towers were gone forever.

In the weeks following, my mind was nowhere close to focusing on law school. I was more concerned about our Nation's response and about what my buddies that were on active duty already were going to be doing as a result of the attacks. After all, I was a Soldier too.

The 9/11 attacks made me want to report for duty at Fort Benning, Georgia for the Infantry Officer Basic Course immediately. At law school, it was difficult to concentrate as news came back about the Central Intelligence Agency's Jawbreaker Teams and United States

Special Operations Forces taking the fight to the enemy in Afghanistan. Of course, things went so well in the beginning that it seemed like the shooting war would be over within a couple of months.

I knew I had earned my time at law school and that I was committed to the Army afterwards, no matter what. So ultimately, I rationalized that my military service obligation would happen whether I left law school early or after I completed law school.

It made sense to stay and finish law school and stick to my original plan – get my law degree, serve the three years I had committed to, and then return to Hollywood to work in entertainment.

Chapter 4: Hollywood

"The key to this business is personal relationships." – Dicky Fox,
Jerry Maguire

October 2001 – December 2002

Building the Business

October 2001 – May 2002

While I was in law school, I was also on the lookout for opportunities to expand my show business horizons.

My long-term goal was to build a foundation in the entertainment industry that would not fall apart during my time in the Army. The details of my plan was to get internships at talent agencies and movie studios, build that into work as a production assistant or agent assistant, then go off to the Army. The business is all about "who you know" and it was very important for me to build relationships. So after my military service, I could return, and secure a job with one of those key contacts.

Aside from that, I was also juggling my client roster which had expanded to include the Pulitzer Prize winner, two Emmy Award winners, and a Tony nominee.

I met the Tony nominee over drinks after seeing *Man of La Mancha* at the Goodspeed Opera House in Connecticut with my oldest brother Patrick. He was actually nominated for a Tony

award for his role in the play *Man of La Mancha*. The Tony
nominee looked a lot like Uncle Phil from *The Fresh Prince of Bel
Air*. He even joked when I met him that he was not Uncle Phil.
He was a courteous man and a talented actor.

During drinks, the Tony nominee told us that he was seeking a new
agent. I told him to look no further. Within days, I signed him.
The Tony nominee then introduced me to two of his friends that
were also seeking representation. Show business was mostly based
off relationships and referrals, so I was interested in meeting them
as well. But more importantly, his two friends had won Emmy
Awards. I immediately signed them too.

And just like that, I had begun building a legit client base.

ABM

October 2001 – May 2002

While at law school, I had also met Joe and Marc. Joe was my first business partner in the entertainment business. He was charismatic and performed magic on the side. His favorite movie was *True Romance*.

Marc became my second partner in the entertainment business. He was well traveled and spoke many languages. He loved driving German sports cars. His favorite actress was Angelina Jolie.

Joe, Marc, and I formed a loose partnership in the way of an informal management company called ABM, mainly to manage the careers of talent. We called it ABM for the initials of our three last names. We all shared a common goal. While attending law school, we wanted to find clients to represent.

We also had the same short-term plan, to obtain interships at agencies and studios across town and leap-frog from place-to-place building contacts. By doing this, we would learn the business better and be able to share different approaches for our future endeavors. It was also so we could gain knowledge on how to successfully run a management company.

All three of us had different strengths and roles that were essential to bringing in the right people and projects. Joe was the "fan" of the group. He enjoyed watching movies and understood what it was that made certain films connect with an audience. Marc was the "internationalist." He had travelled quite a bit (mostly in Europe) and understood the world outside of America – international markets, which were key to indentifying what projects would do well overseas. I was "the insider" by virtue of having been in the business independently for a couple of years before we had met. I was the hook for new prospects.

Our plan after law school was to set the foundation for packaging projects – which means matching writers with directors and stars. We envisioned fully establishing the firm and building upon our client roster and the experience we would gain from working for other entertainment companies. The law degree would also be a big bonus.

With our goal set, it was time to become "seasoned" in the business.

Joe, Marc, and I were spending an awful lot of time working on developing our contacts and building up projects. We added an Academy Award winner to the client roster after following up on a lead from an ad in one of the trade magazines (*Variety* and *The*

Hollywood Reporter, which roughly equate to being the entertainment industry's *New York Times* or *Washington Post*).

Joe and I set up a time to meet this Academy Award, or Oscar winner, at the Beverly Regent Wilshire (the *Pretty Woman* hotel). I went to the meeting completely under-dressed. I wore jeans and a football jersey. I think the staff at the hotel thought I was crazy, due to my being under-dressed. But since we were talking about the movie business, we came across as "industry" and therefore, no further explanation was requested or needed. We were free to drink, dine, and do business.

We were meeting the Oscar winner to talk about three scripts that he wrote and wanted produced. He was in his late seventies or eighties, excited, and passionate about protecting the environment. We found out he had also served as a pilot in World War II. Over the course of our conversation, he told us about his experience in the Far East during the war. He piloted missions against the Japanese. He also discussed some of his early experiences working as crew on several movies since the war. He experienced the Golden Age of Hollywood behind-the-scenes. In spite of his age, Oscar had a lot of energy and passion.

After signing the Oscar winner, we realized that we (3 first-year law school students) had a client roster that included a Pulitzer

Prize winner, two Emmy Award winners, a Tony nominee, and now, an Academy Award winner. Not too bad.

Working With My Mentor

May 2002

Before returning to Los Angeles for law school, one of my old roommates from West Point introduced me to a childhood friend of his that worked in Hollywood. My roommate's childhood friend became my mentor. Over the years, he gave me advice on dealing with clients and possible buyers, as well as giving me career advice in general. We frequently spoke on the phone and exchanged e-mail, but I didn't meet him until I was out in L.A. for law school.

My mentor worked on the lot of Paramount Pictures. His company produced movies at the studio, and for the studio. He was young, ambitious, and blunt. He helped me to understand the business (all the angles) and develop my career. We even tried to collaborate on some projects.

Because he was close friends with one of my good friends, I trusted him. He gave me an unfiltered perspective on the movie business.

In the long-term, I wanted to be either an agent or a producer. He encouraged me to learn more about both being an agent and a producer.

Over the course of my first year at law school, Joe, Marc, and I started packaging an independent film that had a star's interest.

Packaging a film entails bringing together the writer, director, and key talent (actors and actresses) in order to make buyers (producers and/or studios) more interested in taking the risk of purchasing, producing, and distributing the project.

Our client was the writer and director of the movie. He was a Texan, very dedicated to his work, his son, and Texas. He was not a so-called "A List" name, but he had worked on some big projects and he was well-connected in the industry. Due to this, he generated the interest, but not the commitment, of a star. Based on all this info, it was enough for me to take it to a studio executive for consideration.

And so, I took the project to my mentor. My mentor was interested in backing the project. He felt that he could get the financing.

This was exciting. I called the client: "Here's the deal. You will get a $2.5 million for writing the script and directing the film. The studio will commit $20-25 million for the movie. I think you ought to seriously consider this situation."

The client had spent nearly $250,000 out of pocket to get the project off the ground, but he was desperately in need of financing at this point in time.

"I do not want to relinquish creative control," he told me.

"I do not think that they will put money up in that case, but we can ask," I replied.

I called my mentor and asked, "Will they put the money up?" I paused. "And let my client maintain creative control?"

I already knew the answer, but I had to ask.

"No," he replied.

"I figured as much."

I called the client and told him the deal would not work that way. He still did not want to give up any creative control. Had he accepted the deal, his initial payout would have been $2.5 million. Joe, Marc, and I would have gotten $250,000 out of that as our standard ten per cent commission.

The project languished until the spring. I was getting ready to walk into my final exams for the end of my second semester of law school. Ten minutes before my test, I got a call.

It was my client, the same writer-director. "I need $2 million by the end of the day."

My client went on to explain that he had mortgaged his house and tried to get other investors, but they backed out. Now, all the money was gone and the project was about to fall apart.

"Well, let me see what I can do. I have to take a test in a few minutes. Let me make a call," I told him.

I called my mentor again. He and I already had planned to meet up at the end of the day.

"Come on over. We can see what we can do, but he will have to relinquish creative control," my mentor again told me. His openness and willingness to take another swing at the project were both a source of relief and gratitude.

"I think he knows that at this point," I said.

I took my test – Civil Procedure II.

That afternoon, I went to see my mentor at Paramount Pictures.

"We can get the money. The question is whether he will give up creative control," my mentor explained.

I contacted my client again. Before, he was unwilling to give up creative control, and there was no deal. Now, his house was on the line.

"The key point is this – will you give up creative control? We can still make a deal," I told my client.

"I cannot give up creative control," my client told me.

"No way," my mentor said.

A side note here…this deal seems like a "no brainer" to most people outside of Hollywood. In Hollywood, creative control is a thorny issue. In addition, passion runs hotter and faster than common sense, making show business "crazy." These were all things that my mentor understood. I did too, but it was frustrating to lose out on a straightforward deal over creative control in a project that desperately needed outside financing.

When all was said and done, my client nearly lost his house and the movie never got finished.

But my mentor continued to teach me. He saw my potential. He recognized that I understood the entertainment business and that I was good at negotiating and making deals.

"Intern at an agency. Intern at a studio." He told me.

My mentor also wanted me to not only learn the selling, but also the buying side of the business. It pays to be well rounded in this entertainment industry.

The Interview

May 2002

In May 2002, I had secured an interview for an internship at an agency. My interview was to be conducted by a guy who was dubbed the "Big Dog" by one of his fellow agents. I made contact with him by replying to an employment ad in one of the industry trade magazines. The Big Dog was the guy in charge of recruiting new guys like me who wanted to be agents. He was young, ambitious, and very skilled at being an agent at "The Agency".

The Agency was a talent and literary agency that worked primarily with clients in film and television. It was a small boutique agency, consisting of about half a dozen agents and about six interns and assistants. The agents were servicing clients who were mostly supporting cast or recurring cast members on television shows. There were not that many big stars on their client roster. Though they did have several above-the-line clients, mainly writers and producers, who were well-experienced industry veterans with significant credits to their names. But due to the distinguised reputation of its founder and chairman, "The Boss," The Agency had some strong ties in Hollywood.

Going into the interview, I had one previous experience as an intern at another literary management firm. I read screenplays and wrote "coverage" (a cross between a book report and a review) for the firm I previously interned for. I also had three years of experience as a freelance talent representative through my side management firm, ABM. Bottom line being, I had already been working on the margins of show business independently and as an intern, so I was confident in my abilities to fulfill the job duties.

At the onset of my interview with the Big Dog, my goal was to also get in with them, then try and integrate my award winning clients into The Agency's roster.

Separately, and in line with the advice from my mentor, my hope was to land an internship with The Agency first, then follow it up with another internship at a movie studio the following summer.

The Big Dog, on the other hand, had a much different line of thought when I interviewed with him.

"There is going to be a real opportunity for you here," the Big Dog told me at the completion of my interview.

"I look forward to it," I replied.

During my interview, I found out that the Big Dog was both The Boss's assistant and a junior agent with a growing client roster.

After my interview with the Big Dog, he introduced me to The Boss.

The Boss ran the Agency. He was in his early 70s, impeccably groomed, and dressed business casual on most days. His sparkly eyes and intense work ethic demonstrated his youthful outlook on life and business.

As mentioned earlier, he was an "old school" agent who back in the day, had a client roster that included almost everyone, or rather, every "A List" talent from the "Golden Age" of Hollywood. During that time, he had represented dozens and dozens of Hollywood's biggest names – clients who were scattered in pictures on his desk and walls – George Burns, Gracie Allen, Milton Berle, Marilyn Monroe, Elizabeth Taylor, and Ronald Reagan (when he was an actor), just to name a few.

The Boss regaled me with tales from the Hollywood trenches going all the way back to the time when he started out as an agent with MCA under Lew Wasserman. Lew Wasserman was a legendary Hollywood talent agent and studio executive, a very powerful man and well connected pioneer in show business, building his career from the 1940ss and expanding it well into the 90s.

To find out that The Boss had learned from one of the greats in the business, was fascinating. This was the most exciting job interview that I have ever been to in my life.

Before I left, the Big Dog told me, "You got the internship."

I thanked him and left.

Little did I know where the internship would take me.

When I arrived on the first day, I was introduced to the phone, fax, and mail delivery system. My job was to handle all incoming phone calls (there were about two hundred per day), scan and route all faxes (most of them were junk), and screen all the mail (this was very important). All of the pitch letters were sent out with the couriers. Scripts were usually sent out the same way. We received the most important correspondence by courier. We received pitch letters, queries, headshots, and resumes through the regular mail. By screening all incoming mail, I was responsible for determining whether pitches, queries, or acting prospects were worthy of a follow up. Maybe one in twenty pieces of mail was discussed with the agents or The Boss and less than half of those were designated for follow up.

At the same time that our agency was getting pitched, we also were giving several pitches all over town. We were pitching scripts for production, writers for television shows, and talent for movies and television shows. Obviously not all of the clients got the gigs that we pitched. This was the way of "show business."

After my first week, I developed a deeper appreciation for the interning opportunity that I had received. These calls, faxes, mail, e-mail, and our own pitches all vividly demonstrated that the film and television industries are highly competitive.

If you aim to produce, direct, write, or act, then you have about a one to two percent chance to get your foot in the door. From there, you have about a ten percent chance of being heard by a studio. Beyond that, you got another one to two percent chance to get a "green light" and be involved with a project that goes into development. Only about ten percent of the projects that go into development actually get produced. Your odds are slightly better than one-in-a-million to make it in the business. Persistence and resilience are critical to success. These were lessons that reinforced what I learned at film school. Always continue the climb. In this business you can't wait for it to happen, you have to make it happen.

Established (Almost)

May 2002

"There is a real opportunity for you here," The Big Dog repeatedly told me during my first week as an intern at The Agency.

But to be clear, my internship was unpaid. This is the norm in Hollywood. However, the experience and knowledge you gain from an internship at a legit Hollywood agency, production company, or studio is invaluable.

That being said, what I did know was that I needed to make some money. I started job hunting. I told The Boss that I would now be able to spend only one day a week at The Agency. He was fine with that.

During this time, I was also unaware that the Big Dog was going to be out of the office for a couple of weeks for medical reasons. In actuality, he wanted to transition away from being both The Boss's assistant and an agent to exclusively being an agent. The Big Dog felt that going away would accomplish this change. No one knew what was going on. The Boss was surprised by the Big Dog's departure.

Then one day, I was getting ready for a job interview and grabbing a late lunch when my cell phone rang. It was The Boss. This was no less than two days after I told him that I would only be able to spend one day a week at the office as an intern. It had also been only one day since the Big Dog had gone away.

"I need you," The Boss told me.

"I understand, but I can't work for free," I told him.

"No problem. You are hired, you will get paid," he said.

"Okay," I told him. "When do you want me to start?"

"Can you be here later today?" He asked.

"Sure," I said, "Let me just finish up lunch. I will be there in half an hour."

"Sounds good." He quickly hung up.

When I got in, The Boss was going crazy. The phone was ringing off the hook, faxes were flying all over the place, and he was getting frustrated.

"You will be my assistant," he told me.

The Boss then explained what his expectations were. I was to tend to the phones, email correspondence, client appointments, producer and studio appointments. In my downtime, I could develop my own projects, under his close supervision.

The Boss once again told me about how his mentor – Lew Wasserman – ran MCA-Universal for decades. I'd learned that Lew Wasserman basically created the star system. MCA dominated Hollywood for years, representing the biggest names in show business including Bette Davis and Marlon Brando.

The Boss told me that he worked in the mailroom at MCA out of high school, and that one day, Lew Wasserman called him in.

The Boss went on to tell me the story of how he – The Boss – became an agent…

"I need you to take care of one of our clients," Lew Wasserman told him.

"You got it," The Boss told him.

"You will take care of George Burns," Lew Wasserman said.

And just like that, The Boss and George Burns came to be agent and client. He had a picture of George Burns up on the wall in his office.

He would often tell people stories about George, and at first, I got confused and thought he was talking about me. Usually, after about two seconds, I realized, he was talking about George Burns and not me.

The Boss said he used to have dinner every Thursday night with George Burns and Gracie Allen for several years. Many of the "George" stories that The Boss amused clients with were based upon those dinners. He respected and loved George Burns and Gracie Allen. I know The Boss felt he was lucky in life and no doubt felt a large part of that had to do with George Burns being his first client and staying with The Boss as his agent for decades. I was always delighted or fascinated by The Boss's stories.

In between those, there was also a lot of work to do. I was back to reading scripts, writing coverage (script reviews), reviewing the trades, screening pitches and headshots, handling clients, and setting up meetings for the boss. Only now, I was a fulltime employee with a future at The Agency. I learned a ton. I learned about the history of Hollywood. I learned about deal making. And I learned about the good, the bad, and the ugly side of show business.

Being at arm's length from The Boss for ten hours a day was like going from grade school to grad school overnight in terms of learning the ropes.

In the Golden Age, clients were identified by scouting, primarily at theaters and clubs. The industry invested more in talent, because under the studio system of long-term contracts, there was more at stake for a studio to lose when talent was unavailable. Deal making had not changed much. The leverage had shifted from the buyers – producers and studios to sellers – agents and other representatives of talent, because everyone is a "free agent" in the marketplace of making movies.

I started going out to events at nights. I went to theaters, comedy clubs, and bars. My purpose was to scout new prospects or talent – writers, actors, actresses, etc.

"Funny is always good," The Boss would say. "People need to laugh."

That was a lesson he said he learned with George Burns.

Most of the prospects were not ready and it was my job to report that to The Boss. It was also my job to break the news to the people, that we were passing on them. This was not always easy to do, because some of these people had dedicated a lot of time, money, and energy in their endeavors. In spite of that, I felt that these talent prospects needed to know whether there was bona fide interest.

For instance, a writer would send in a script. The writer would call and ask if we received the script or call to ask what we thought about it. The "Industry Standard" was to say, "We're thinking about it" or "It's under consideration."

This would give those aspiring prospects hope.

I disagreed with the industry standard. If there was no interest, then I usually told the prospects. It was the best thing to do and the right way to treat these people.

I also was in charge of hiring interns. I decided to call my friends, Joe and Marc and get them on board. I figured what a better way for us all to get the best out of the experience.

Within three weeks of being hired, I put my negotiating skills to the test and negotiated an agreement with The Boss. That the interns and the other non-agent staff were allowed to develop their own leads and work their own deals, with The Boss's permission and supervision. Any deals that closed would entitle that young dealmaker to a commission.

In addition to that deal, I had asked for some of us, including myself, to become junior agents. The Boss agreed.

To put this in context, I originally planned to become an agent within three years of returning to Hollywood after my time in the Army. It was part of my ten-year plan. But here I was, three weeks into my internship and already promoted to Agent.

It had already happened. I was now a newly minted talent agent. I was nine years ahead of schedule. It was what I had hoped and planned for. For now, everything else seemed to take a backseat to this.

The next step now was to close a deal – make my first sale as a Hollywood Agent.

Hollywood Agent

May – August 2002

By now, I brought in all of the folks that I represented on a freelance basis and started repping them through The Agency. I also started working as hard as I could to get their projects moving from concept to script to option to screen.

At this point, I also resurrected one of my film school projects. Back at film school, we had to write a film treatment and production plan for a motion picture. I chose to do a biographical picture based on the life of the singer Bobby Darin.

I later showed the Bobby Darin project to The Boss.

"You know," he said, "George Burns was pretty close with Bobby Darin."

"I know," I told him.

"I also know Bobby Darin's old manager," he said.

"Really?"

"Really."

"Let's give him a call."

We did. The bottom-line of the conversation is that the movie *Beyond the Sea* was already in development and the Bobby Darin "story" was essentially in lockdown as a result.

"At least we tried," The Boss said.

I thanked him.

We went back to work.

I was still working hard on making deal number one – my first score.

Day-to-day, I would be setting up meetings for The Boss and managing his relationships with clients and producers. We reviewed new projects and stoked old projects for clients, including packaging top talent.

Normally, I would arrive at the office around nine in the morning and leave around eight at night. Same went for The Boss, who had incredible stamina. Many times, the day ended with me getting him on the phone with his wife so that he could tell her his usual: "I am rounding third and heading home."

After work, it was back to the same routine – meeting with prospects at comedy clubs and shows, scouting for new talent.

I was eating, drinking, and breathing show business.

An Aspiring Actress

July 2002

Hollywood is basically a ghost town around the 4th of July. A couple of weeks before the holiday, I kept getting a call at the office from an aspiring actress. She was very persistent. She called every day looking for representation.

"But they told me maybe," the aspiring actress said.

"I understand," I replied.

"Maybe" in Hollywood-speak meant that there's no deal for you here. Of course, the aspiring actress had no idea about that.

I told her to come by the office the day after the 4th of July. I knew that it would be slow. I knew that I would have time to meet with her.

The aspiring actress showed up early.

"What have you got?" I asked.

She handed me her headshot and resume. She was pretty, but she had very little experience.

"Here's the real deal," I told her. "You don't have enough experience for this agency to pick you up. Straight up."

She was heart-broken.

"But they said maybe," she told me again.

"I know," I said. "But they're just being nice. I'm telling you the truth."

I then told her what she needed to do to get more experience in order to have a better shot at getting representation.

She thanked me.

I felt that the least I could do was tell her the truth and explain the realities of the entertainment business. It's not an easy business to be in. People will tell you right off the bat if you have "it" (meaning talent) or if they like "you" (meaning your look as an actor/actress or your voice as a writer). In the event that you do not have "it," then you will hear a litany of compliments (but no commitments).

The aspiring actress was too young and inexperienced for me to bring her in as a client. She didn't know that. But after we met, she had a better idea of what needed to be done in order to get more experience under her belt to attract an agent.

I believe everyone has a talent, ability, or skill that he/she can mine to support themselves and succeed in life.

However, the reality of show business is that it's cutthroat, pure and simple. In addition to looks or talent, persistence and dedication to actually learning the craft plays a key factor in breaking in.

Elizabeth Taylor

July 2002

From time to time at The Agency, The Boss would captivate me with one of his tales from his early days in the business.

One day, The Boss begins to tell me a story...

"I get a knock on my door. My boss comes in. 'We have a problem,' he says. 'Elizabeth Taylor does not have a date for the Oscars,'" The Boss explained.

"So, then what happened?" I asked.

"My boss asked me to take her," The Boss said.

"And?" I anxiously asked.

"And my wife, at the time said, 'Over my dead body.' So, I almost got to go the Oscars with Elizabeth Taylor."

"That is unbelievable," I said.

It was truly amazing to hear him tell these stories. It was definitely one of the perks of the job. But aside from being entertaining, his stories also taught me a valuable thing. This man had not only been an agent since the "Golden Age" of Hollywood and rubbed elbows with movie stars and tinsel town royalty... but he had also been their friend and had the fortunate opportunity to really get to know these historic and legendary greats in the business.

Consequently, I learned that to promote a positive creative environment, it was just as important for me to become not only their agent, but their friend as well.

Streaming Content

July 2002

In July 2002, The Boss was developing an intriguing model for distribution with a cell phone company. Due to my non-disclosure agreements, I can't provide the names of the key parties. Needless to say, the cell phone maker had developed a technology that enabled content to stream in color through its devices. The resolution was pretty good.

At the time, cell phone networks were expanding capacity. We developed a concept to stream television-like content over the wireless network and on the new phones. The Boss was very excited. But we needed content.

We started talking with producers and distributors about streaming their content. However, there was a fair amount of skepticism.

We were told: "No one wants to watch television on their cell phone."

We were asked: "Do you really think this technology will work?"

And were told: "I do not think that wireless networks will be able to handle that much content."

And on and on and on…Needless to say, The Agency was not able to close a deal for the cell phone maker.

In retrospect, this event was amusing. Not too many years later, in 2007, Apple released its first iPhone. The iPhone brought content – music, television, movies – to the consumer on a handheld device, followed by a host of other companies. Streaming live television and downloading television shows and movies from the Internet are now part of the mainstream culture, distribution, and exhibition ecosystem for show business.

Much of the entertainment business is based on instinct. No one could predict that cell phones would become so accepted in viewing content, just like no one can predict what movies will be a hit. But what I learned was that just like finding those golden people to become profitable stars, finding profitable opportunities is also what a good agent keeps an eye out for.

The Backdoor

August 2002

In August 2002, I was also exploring ways to sign on new talent for The Agency by doing it the old-fashioned way – scouting. We received several invitations each week to attend live concerts, plays, comedy clubs, readings. You name it. Almost every other night, I went out to a bar or club to see a singer, comic, or actor.

While scouting did not yield many clients, it did yield some. I also found that clients that we met while scouting were pretty dedicated. I guess that stemmed from the notion of setting the foundation for the client-representative relationship. The gist of it being— you (talent representative) came out to see me (prospect), you (talent representative) liked me (as a client), and I (talent) liked you (as a representative).

One night, Marc and I were headed to check out an act at one of Hollywood's most famous comedy clubs. Marc had forgotten his driver's license and they were not going to let us in. I told the bouncer at the club that we were invited.

The bouncer asked us, "So, are you industry?"

I replied, "Yes, we are. We're from an agency."

The bouncer stepped aside, "In that case, come right in."

The staff picked up a table, led us to the backdoor, walked us through the club, and set us up in the front row. We received complimentary drinks and appetizers, and we didn't pay cover.

It was reminiscent of the scene in *GoodFellas* where Ray Liotta's character Henry Hill and girlfriend are taken through the nightclub only to be front and center for the show. That same scene was paid homage to by Vince Vaughn and Jon Favreau in *Swingers* a few years later.

I don't remember whom we were scouting that night, but I will always remember the *GoodFellas / Swingers* treatment. The VIP experience was one of the nicer perks of being "in the business."

The Pink Slip

August 2002

Not long after I had my foot in the door of the Hollywood establishment, I received bad news from law school. My school, Southwestern University School of Law, sent me a letter that basically told me law school was over. I was no longer admitted as a student there. The same letter also stated that I could appeal. I appealed. I lost. The bottom-line was that my grades were not good enough to continue law school.

As part of my previous ROTC agreement with the military, it was lawfully mandatory for me to send a letter to the Army informing them of this change. I wrote the Army and reported that I was not continuing law school. Which meant for certain, I would be required to report for active duty. I enclosed my letters from school, as well as my commissioning papers from the Cramer's Sabers Battalion. Upon my commissioning, I was assigned to the Infantry with follow-on duty in the Judge Advocate General's Corps (JAG). Since I was now going to be reporting for active duty in the Army, this was the way that I wanted to go on duty.

Emotionally, I was not ready to leave show business. Mentally, I knew that it was my duty to do so.

The Army replied to my letter and advised that per the binding mandatory agreement, I would now be ordered to active duty, but that I also had to be re-accessed.

'You have to be kidding me,' I thought. I was not too happy, when I read the Army's reply about being "re-accessed."

Accessions is the name the Army provides for the tedious administrative process that places newly commissioned officers into a specific branch within the service, be it from West Point, ROTC, or the Officer Candidate School (OCS). It entails a ton of paperwork and approximately six months to go from application to branch assignment and initial duty orders. Being re-accessed meant that I had to go through accessions all over again.

This gave me six months to prepare. I now had six months to get back into respectable shape for the Army. It also meant that I had six months to live, work, and exist before I officially reported for active duty.

If I had to leave Los Angeles, The Agency, and the entertainment business, then I wanted at least to have the job I wanted in the Army. However, I was stuck with six months of waiting, preparing, and despairing. The Infantry was at best on hold. Show business was soon to be an object in my rearview mirror.

I had a lot of time on my hands that I was not anticipating; this was truly like being in a state of limbo. I was in limbo with the Army regarding what my future military job would be. I was in limbo with my life, because I was waiting on the Army to complete the accessions process. I was in limbo with show business, because even though the Army had given me more time to be in Los Angeles, I couldn't bring in new prospects. I was in limbo with being an agent, as I would soon be reporting for active duty and becoming a soldier.

It was time to have a drink.

I called up the Big Dog. He had come off his hiatus and gotten back into agenting. We went to happy hour at the Stage Deli across the street from The Agency. I needed a drink (or two or three) very badly. The Big Dog was more than willing to oblige. He was the one that brought me into The Agency. He was the one that initially hired me at the Agency, and set me up to become The Boss's executive assistant, and thus, becoming an agent. He helped me develop my abilities as a rep to writers, directors, producers, and actors. The Big Dog had played a key role in making my path to show business success a real possibility. Now, it was over.

I was out of time, at least for the next three years, and had to report for duty. At the age of 23, three years seems like a very long time.

More importantly, I was going from a Hollywood focus in the business of being an agent to a military focus in the business of becoming a platoon leader in the Army. These were two very different lenses through which to view the world.

"I can not believe it," the Big Dog said of my news.

"I knew this day would come, but I never expected it to come this soon," I told him.

"You have to go?"

"Yes, I could face serious repercussions if I don't." I continued, "It was part of my initial ROTC agreement. I have to honor that".

"I'll drink to that!" He said.

We ordered another round.

The Big Dog respected that I was going into the Army, but he didn't have any military experience, so most of what I was heading to was foreign to him. He was eager to get back into the mix of Hollywood. So was I, but the army was now waiting for me.

Preparing For Duty

" Seems it never rains in Southern California / Seems I've often heard that kind of talk before / It never rains in California / But, Girl don't they warn ya / It pours man it pours" – Albert Hammond, *"It Never Rains In Southern California"*

August – December 2002

I spent most of the fall getting myself back into a respectable level of physical fitness.

Being a new talent and literary agent in search of deals, I had somewhat lived the life of a semi-vampire. I was in the office during the day and visiting comedy clubs and cocktail mixers at night.

But I wound down my affairs in the entertainment business as an agent, increased my physical fitness regimen, and sent my papers into the Army to see where they would send me.

Winding down my affairs in the entertainment world was very difficult. I knew that I had to leave Hollywood, but I had gotten so close to my goal, being an agent and all, that it was hard to have to leave. I procrastinated until there was almost no time left.

One morning in December of 2002, I popped the top off a bottle of Jack Daniel's, poured a tall glass with ice and began writing all of my termination letters. I was terminating all of my representation agreements as an agent with my clients. I was terminating my ties with Hollywood. It felt like I was cutting off one of my arms. A couple of hours later, all of the letters were written. I put stamps on the letters and sent them away.

I told Joe and Marc that I was out. We had forged the foundation of a representation enterprise that we envisioned would eventually give us our livelihoods. Now, our partnership was over, and they would proceed on their own.

I remember telling my Pulitzer winning client that it was fun while it lasted.

I reached out to the Emmys, the Tony nominee, and Oscar clients and let them know that my time in the entertainment business had reached its end. One by one, I told the rest of my clients the same.

I called my mentor on the Paramount Pictures lot. We met up and had drinks.

"Well, until the next time," he said.

"You bet," I told him.

Aside from being upset at having to leave my job as an agent, I was pretty upset that I was also not guaranteed a return to the Infantry; my original pick from the time that I got commissioned.

Sometime, between Halloween and Thanksgiving, I got my letter from the Army. It said I was going to stay in the Infantry. That was good news. Then, I had to report for duty at Fort Benning, Georgia in order to attend the Infantry Officer Basic Course in January.

Chapter 5: The Army

"But the most important lesson I took away from those early days of combat in Iraq, was the primacy of the 3Ms" (the Mission, the Men, and Me). – Pete Blaber, The Mission, the Men, and Me

January 2003 – January 2005

On the Road

January – August 2003

My best friend from back home, Emmanuel, came out to take the trip from Los Angeles, California to Fort Benning, Georgia. It was my first cross-country road trip, but not his. He traveled cross-country during his college years, between his transferring from Wheaton in Massachusetts to Rutgers in New Jersey.

On his first cross-country trip, Emmanuel drove from Massachusetts up to Northern California – getting to see the Redwoods – then, he returned to the East Coast.

I was grateful that Emmanuel came out to take the trip from L.A. to Fort Benning with me. We went to high school together, wrote and performed in *Good Mariners* together, and hung out often. We used to play a lot of wiffle ball, when we hung out. He visited me at West Point and I visited him at Wheaton and Rutgers. He was there when I became an Army officer and graduated from college. I was also at his graduation from Rutgers. It was good to have my best friend on what proved to be a fun trip, but also a difficult time for me (leaving Los Angeles).

Emmanuel and I made our first stop in Las Vegas, Nevada, an oasis of 24/7 alcohol and gambling. In Vegas, we met up with another friend. We danced, drank, and gambled.

Next, Emmanuel and I drove through Phoenix, Arizona. The desert mountains of Arizona were beautiful. We saw the Hoover Dam. We continued on and stopped in Las Cruces, New Mexico. We gazed at stars in the New Mexican night sky. We then drove through western Texas on Interstate 10. This was quite an experience. At one point, there was something like 200 miles between fueling stations and rest areas (or at least it seemed like it).

Eventually, we passed by the Alamo and arrived at the River Walk in San Antonio, Texas. Emmanuel and I met up with my older brother Matt. Matt was in town on business, but it was also Super Bowl Sunday. We watched the game in San Antonio.

Matt, Emmanuel, and I were mostly focused on the Super Bowl. I think Matt and Emmanuel understood that I needed a diversion from both Hollywood and the Army. I was really drained. Leaving Hollywood took a toll on my psyche. Having to wait so long for the Army to tell that I was still going to be in the Infantry also took a toll. I guess more than anything else, I was just tired.

After the Super Bowl, we went to some of the bars on the River Walk, talked about Matt's trip in Texas, and the trip Emmanuel and I were taking to Fort Benning, Georgia.

From San Antonio, Emmanuel and I traveled to New Orleans, Louisiana. We had a lot of fun in the French Quarter and Jackson Square. We had beignets at Café Du Monde and plenty of great food.

Emmanuel and I stopped in Biloxi, Mississippi as a waypoint on our last leg of the trip. We both loved the movie *Biloxi Blues* and the stop reminded us of this. We had Po Boys on the Gulf of Mexico while in Biloxi.

Finally, we wound our way to the infamous Victory Drive of Columbus, Georgia. Victory Drive, as is the case with almost all strips leading to active Army installations, was replete with strip clubs, pawnshops, payday lenders, and used car dealerships. Full depression sank in. I was a long way from sunny Southern California.

"Should I take your razors?" Emmanuel asked me. He was half-joking and half-serious.

"No, I'll be fine," I told him.

For me, returning to Fort Benning, Georgia was bittersweet.

Emotionally, I had still not let go of the fact that I was done in
Hollywood.

Mentally, I was eager to start my training.

The next day, I reported for my first day of training at the Infantry
Officer Basic Course at Fort Benning, Georgia. I was at Fort
Benning from January – August 2003. I drank a lot during that
winter and spring. We used to go to a place called The Tap
frequently. But during summer, I quit drinking. I went "cold
turkey" in order to focus on my next assignment – being a platoon
leader and getting ready to go to war.

By summer, I was ready to proceed to Fort Hood, Texas and begin
my first assignment. I learned that I was going to the 1st Cavalry
Division. The unit was preparing to go to Baghdad, Iraq. I got
ready to take charge of a platoon of America's finest Soldiers.

I showed up at Fort Hood on a Saturday and figured that I would
be fine in my California garb and relatively long hair since it was
the weekend. Boy, was I wrong. The young Soldier at the CQ (in-
charge of quarters) desk told me that the battalion Executive
Officer was in; the 2nd in command of the battalion; my new

boss's boss. By the way, I had not even met or spoken with my new boss,who was my company commander, at this particular moment in time. The battalion Executive Officer was a 'by the book,' firm but fair, and very Army kind of guy. He was a fit, solid, and skilled leader. I went in, saluted him, and got an overview of the battalion.

The Executive Officer looked at me as if I was from outer space. The usual accepted way of reporting for duty as an Infantry platoon leader is to get a "high-and-tight" haircut, where the sides of your head are shaved and there is stubble on top, and be wearing your uniform. I was far from that. I was in civilian clothes, had a haircut like Elvis Presley, and I was a little worn from the cross-country trip.

I walked in, rendered a hand salute, and said, "Sir, Lieutenant Mannix reporting as ordered."

The Executive Officer gave me the 'one over' from head to toe and had some questions.

"Where did you come from?" He asked.

"Los Angeles, California," I replied.

"I see," he said, handing me a book and some papers. "Welcome to the battalion. Here is the Unit History, a book that the commander wants you all to read, and the training schedule."

"Thank you, sir", I replied.

I was then told that I would take over a platoon during my first week on duty.

"PT Formation is 0630 Monday. You will meet your platoon then," The Executive Officer advised.

PT meant physical training. I was also informed that in about six months, we were going to war – Baghdad, Iraq.

"Roger, sir," I acknowledged and rendered a salute.

The Executive Officer returned my salute. "Dismissed."

The Reapers

August – December 2003

I had now fully transitioned into army life. My formal title was 1st Lieutenant Mannix – Platoon Leader of 1st Platoon, Alpha Company, 1-5 CAV. My Platoon, 1st Platoon, was called "The Reapers." That name was chosen by the Platoon Sergeant – Woody. Woody had served in the platoon for several years during his career. He had spent his entire career "on the line" meaning in units either in combat or preparing for combat. His experience was invaluable. Over the course of his career, he had served in every single position within the platoon – rifleman, machine gunner, grenadier, team leader, squad leader, driver, gunner, vehicle commander, and section leader. He cared about the things that mattered most – doing the job well and staying safe.

Furthermore, as the platoon sergeant, he was the heart of the platoon – the most experienced Soldier and the 2nd in command, after me – the only officer. Our company, Alpha Company was the "The Annihilators." I reported directly to the company commander. Our Battalion, 1-5 CAV, was called "The Black Knights."

We learned in September that we would not deploy as 1-5 CAV, but instead our company would be assigned to 2-12 CAV, called "Task Force Thunder." The other units in Task Force Thunder were tank units; our company was the only Infantry unit assigned.

My platoon was a Mechanized Infantry Platoon. The platoon consisted of two sections that included two Bradley Fighting Vehicles (BFV) per section (including a driver, gunner, and vehicle commander for each vehicle), and three squads consisting of nine Soldiers. The Bradley Infantry Fighting Vehicle has three weapon systems – a 25mm main gun and a 7.62mm coaxial machine gun that is designed to destroy enemy Infantry, as well as tube-launched, optically tracked, wire-guided (TOW) missiles designed to destroy enemy armor. Each squad consisted of two fire teams. The fire teams included a team leader, grenadier, automatic rifleman (machine gunner), and a rifleman. Each squad had its own leader.

The primary operations concept behind the Mechanized Infantry Platoon was the ability to maneuver rapidly, mass fires, and seize terrain. The Bradleys were armored. They could sustain direct and indirect fire from several weapons. This aspect of the Bradleys made them an ideal vehicle for conducting mounted patrols in Baghdad, because they could sustain strikes from several types of Improvised Explosive Devices (IEDs).

Figure from Page 1-6, ATTP 3-21.17, *Mechanized Infantry Platoon and Squad (Bradley)*, November 2010.

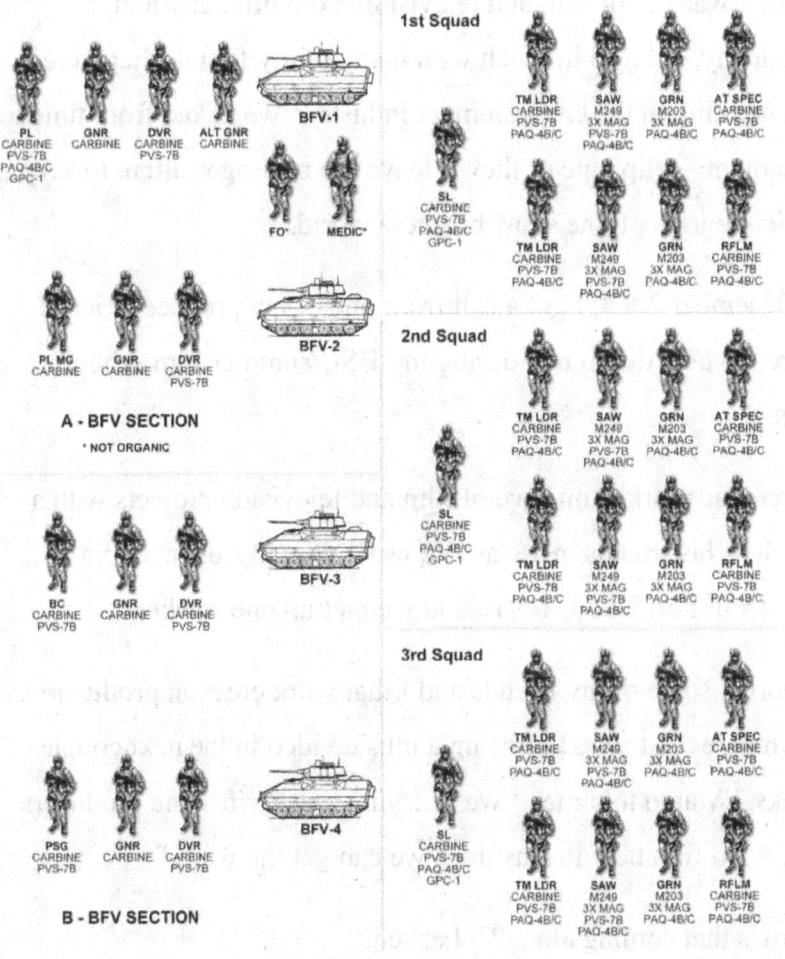

Rewind: Merry Christmas

December 2004

While I was out of film and television, both officially and practically, I stayed in touch with many of my friends that were still working in the entertainment industry. We talked from time to time on my cellphone or they'd leave me messages often, to keep me in the loop of the show business world.

In December 2004, I got a call from one of my producer friends, Marco, who I'd also met during my USC summer film school days.

Marco had worked on several film and television projects with a bunch of his friends in Texas. He was living in Austin, about an hour from Fort Hood, Texas, and we met up one weekend.

"George, some of my friends and I that work crew on productions down here are trying to line up a music video in the next couple of weeks," Marco told me. "We're trying to convince the producers in L.A. to film here in Austin so we can get the work."

"How's that coming along?" I asked.

"We could use some help," he told me.

"What kind of help?" I asked.

"Well, since you are asking…"

Marco went on to tell me that he wanted me to call the L.A. folks and convince them that Austin was cheaper and "cooler" for this particular video. I helped Marco put together the music video deal.

A week later, Marco called, "Thank you, George. They are going to film in Austin, before Christmas."

"Glad to hear that," I told Marco.

I found out later that they almost had to cancel the shoot. The original plan was to film on the campus of the University of Texas – Austin. Unfortunately, the production schedule overlapped with final exams. The University did not want a production crew in the middle of campus during finals. The shoot was moved to the downtown area in Austin, where the 20-something crowd goes barhopping.

Months later, while I was in Iraq, I checked back on the video for Ryan Cabrera's song "On the Way Down". It was number one on the Billboard charts and the video was playing widely on MTV and VH1.

This Is Real

January 2004

Fortunately, I got to spend four months training the Reapers for our upcoming deployment. This also enabled me to get to know the Soldiers and for them to get to know me better. By the time we left Fort Hood, Texas to conduct combat operations in Iraq, we were a cohesive, well-trained, motivated platoon.

Upon arriving in Baghdad, Iraq, we were based at Camp Victory North. Camp Victory North was very close to the Baghdad International Airport. Our area of operations was the eastern part of Abu Ghraib on the western outskirts of Baghdad.

As we departed the base, I often told the men in my Bradley, "Good morning, gentlemen, welcome onboard the Jenny. This has been designated a non-smoking ride by the United States Government. Weather today is expected to be – hot!" I named my Bradley Fighting Vehicle "the Jenny" in honor of Jenny McCarthy. I thought Jenny McCarthy was beautiful and figured that pictures of her smiling in a bikini would motivate us throughout the tour of duty in Baghdad.

One January morning, we were conducting an improvised explosive device (IED) sweep of the major roads in our task force area. Since it was only our second day, we were following the unit that had been patrolling the area for the past several months. My counterpart, the platoon leader from the unit that had patrolled the sector for the past several months, led the patrol.

My platoon and I were briefed prior to deploying that the marketplace – an open air "bazaar" in the middle of town – was "hot" with a lot of enemy activity. There were several IED attacks and small arms fire with insurgents and terrorists in and around the market.

As we approached the market on this dawn patrol, a blast of dirt came up along the side of the Bradley that I was following.

"I-E-D at 9 o'clock," my counterpart radioed. "Two whiskey-India-alpha."

Two whiskey-India-alpha meant two Soldiers wounded in action.

Both my counterpart and his gunner sustained injuries. Fortunately, their Bradley armored vehicle took the brunt of the blow and my counterpart and his gunner only sustained minor injuries. We returned to their base camp. The medics treated my counterpart and his gunner. Within fifteen minutes, they mounted back up to lead the patrol and complete the mission to search for IEDs.

For my counterpart, this was one of his last engagements with the bad guys, at least on that tour of duty. My counterpart and his Soldiers were scheduled to rotate back home very soon. For my Soldiers and I, it was our first contact. I was shocked, inspired, and hardened by that incident and I believe that the event had a similar effect on my Soldiers.

We realized, this was war, and people could die. We would have to kill the enemy or else the enemy would kill us. It was that plain and simple.

After completing the patrol, I spoke with my counterpart. The purpose of us patrolling with the other platoon was to learn about the area of operations and the enemy. My counterpart had been in Iraq for quite a while and was obviously more experienced with the area.

"How come you didn't have one of the other units complete the patrol after you got hit?" I asked.

"They need rest. We know what we're doing. This was no big deal today." He told me.

I didn't say anything back.

My counterpart continued, "We have to take this. We have the armor. We have the firepower. If we do not do this, then some light-skinned vehicles will get blown away."

At first, I thought he was a little crazy. 'We have to take this? I don't think so,' I thought.

It took me a while to realize fully what my counterpart was saying. The bottom-line is that we are Mechanized Infantry, and that means that we have more armor, more firepower, and more skills with which to engage the enemy. It also means that, if and the key word is "if" here, there is an engagement, we would be the best suited element to take on the enemy. Eventually, I would come to agree with his thought process fully.

In the meantime, I reviewed the action with my Soldiers. We needed to be vigilant. Vigilance was critical to success. It meant finding the enemy before the enemy found us.

We Are SWAT

January – April 2004

The Reapers effectively executed vigilance. We, the Reapers, were able to find the enemy on numerous occasions. Normally, the enemy surrendered without resisting our force. We were recognized as successful.

We became the "go to" team for our task force, due to our success in the field.

If the Task Force Headquarters said, "We got a high-value target. We need this guy captured." Then we, the Reapers, got the call and we executed the mission.

If the Task Force Headquarters said, "We need a detainee escorted," the Reapers got the call and we executed the mission.

If the Task Force Headquarters said, "We need a site secured," this normally meant that there was a suspected roadside bomb or one just went off. The Reapers would get the call, and then we would go out to secure the site.

I told the men that we were the SWAT team for our task force. The way we conducted operations was pretty simple. In the daytime, we walked around the area and met the people. At nighttime, we hunted the enemy. My platoon sergeant, Woody, would go out and set up observation posts and I would take out a patrol team. Basically, the observation posts were static sites from which my platoon sergeant and his team would look and listen for the enemy.

The intent is to see the enemy before they see you in order to take action first. We normally occupied a house and set up the observation post on the top floor or roof. We usually gave food, water, and candy to the family that owned the house to make them feel comfortable, or at least a bit less scared.

My patrol element was on the move, but usually in and around the static sites. Our plan, which normally worked quite well, was to have the observation posts identify suspicious activities or people and then let the patrol drive in or walk in and take action. By doing this, we were able to capture a lot of known and suspected enemies. We were very good at what we did. We were also lucky.

However, luck does not last forever.

Our luck almost ran out in late February of 2004. We were tasked with escorting a fuel truck around the task force area of operations to provide re-fueling services to other units.

In the middle of our mission, we were also asked to execute some other escort activities that delayed the re-fueling timeline. This was bad. We normally wanted to get the fuel truck around the town as fast as possible, because the fuel truck was loaded with flammables and was a soft target, meaning it had minimal armor and very little firepower. On this particular day, night fell, and we still had to get the fuel truck back to base.

I was leading the patrol. We were on an escort mission. I was monitoring message traffic from our higher headquarters, Task Force Thunder, via the radio for information about recent enemy activity.

"Thunder Mike, this is Annihilator Red One, over," I radioed.

"Annihilator Red One, this is Thunder Mike, go ahead, over," Task Force Headquarters replied.

"Thunder Mike, this is Annihilator Red One, departing rally point in order to bring the fuel trucks back to Camp Victory, over."

"Annihilator Red One, this is Thunder Mike, roger, out."

We finally pulled the convoy out of the gates.

I was also speaking with my gunner about getting back to base and getting some rest. We had been "outside the wire" (or beyond the relatively safe confines of base camp) for over thirty hours. Less than four hundred meters into our movement, the fuel truck was hit with an IED. The fuel urtank burst into flames.

Two vehicles moved in a suspicious manner past our convoy. I engaged the cars with my rifle.

My gunner identified personnel observing the blast area on the other side of the road. I assessed that the observers were acting as spotters for the bombers. I directed my gunner to fire at them.

"Annihilator Red One, this is Thunder Mike, over."

"Thunder Mike, this is Annihilator Red One."

"Annihilator Red One, we are receiving reports of shots fired, over."

"Thunder Mike, that is affirmative. The fuel truck was hit with an India-Echo-Delta (IED or Improvised Explosive Device, a roadside bomb), we engaged, break," I continued. "Enemy appears neutralized, preparing to dismount, over."

The Soldiers in my wingman's track were already on the ground. My gunner had our driver drop the door and now the Soldiers were getting out. I told my gunner he had the Bradley vehicle to himself. I was getting out so I could lead the dismounted Soldiers on foot.

I got onto the ground and led the dismounts to the burning fuel truck. One of my Sergeants from my wingman's track stopped us en route. He advised that no recovery was necessary, because the driver and passenger escaped from the fuel truck with only minor injuries. By this point, other units from the task force arrived on scene to assist.

On the ground, I linked up with the leaders of these other units and reviewed the situation. As we were assessing the situation, rounds inside the fuel truck started to go off.

We quickly realized that the rounds were "cooking off" due to the fire.

"Oh, yeah, they forgot to get their weapons on the way out of the truck, sir," my sergeant told me.

"Well, that's understandable," I told him.

The radio cut me off. "Annihilator Red One, this is Thunder Mike."

"Thunder Mike, this is Annihilator Red One."

"Annihilator Red One, this is Thunder Mike. We are receiving reports of shots fired in your area."

"Thunder Mike, Annihilator Red One, that is affirmative. Rounds are cooking off from inside the fuel truck, over."

"Annihilator Red One, this is Thunder Mike, roger, out."

With that, I led a search team down some of the alleys in order to see if there were any suspicious activities, people, or other concerns. We did not find any additional information or enemy in the alleys.

A few weeks after this firefight, my platoon was securing our patrol base. Shortly after night fell, I sent Woody out with one of the squads to set up observation posts in the neighborhood. After they got into position, I took out another squad on a mounted patrol. We usually had two Bradleys and one Humvee out on patrols. I left one squad behind at the patrol base to pull security there.

About fifteen minutes into my patrol, I observed some suspicious activity. I decided to dismount with my squad and check things out. As we started moving through the narrow street, we came under attack from both sides of the street. Rifle and machine gunfire started coming at us from the roofs.

During the initial burst of gunfire, part of either a round or the wall behind me bounced off my sleeve. Miraculously, it did not break my skin.

My Soldiers and I returned fire quickly and silenced the enemy's guns.

"Red One, Red One, this is Red Four, you OK? Over," Woody radioed.

"Red Four, Red One, roger. Engaging enemy presently, over."

There was a silence. Then, Woody's observation posts came under fire.

"Red Four, Red Four, this is Red One, status, over," I radioed.

"Red One, this is Red Four, receiving small arms fire. Returning fire, over," Woody replied.

"Red Four, this is Red One, roger, out."

As I received Woody's status via the radio, the patrol base began taking mortar fire.

"No injuries. We cannot see any enemy," the patrol base reported via radio.

I acknowledged the report.

"Red One, this is Red Four, enemy neutralized at our position, over," Woody radioed.

"Red Four, this is Red One, good copy. We will be searching houses for enemy, over," I replied.

"Red One, this is Red Four, roger."

Neither Woody nor the Soldiers with him nor my element were able to locate the enemy in the houses we searched. After finishing our searches, we identified one enemy that was killed in action.

"Thunder Mike, this is Annihilator Red One. We have one enemy kilo-India-alpha (killed in action), over," I reported to the Task Force Headquarters.

"Annihilator Red One, this is Thunder Mike, standby at your present location for Iraqi police, over."

I acknowledged their request. The Task Force Headquarters advised that the Iraqi police were en route to take the body to the morgue.

When the Iraqi police arrived, they had a pick-up truck. They threw the body into the truck and took off.

"That's how they do that?," I asked Woody.

"Well, I guess they see things differently here," he said.

"I guess so," I replied.

As I saw it, Woody was a top-notch non-commissioned officer. He had always been with Soldiers. As I said before, he served in every single position in the platoon. He was also a combat veteran from the first Gulf War and he had served in the Balkans too.

Woody was a Soldier's Soldier. He was a lot like Willem Dafoe's character in *Platoon*. Having him as my deputy was a blessing. We worked together very well.

Holy Week

April 2004

In March 2004, Moqtada Al Sadr was another loud mouth in Baghdad, Iraq. This meant that he was not someone to worry about. There were enough terrorists, insurgents, and criminals from many different groups willing to fight Americans. Then, the Coalition put Al Sadr's newspaper and political offices on its list of places to raid, and shortly after this, things really changed.

April 4, 2004 (Palm Sunday)

In Sadr City, which was on the other side of town from my platoon's sector, all hell broke loose. Four police stations were seized and several Americans were wounded and killed going back into Sadr City fighting to take them back. News reports said that eight Americans were killed and thirty were wounded while fighting to retake the police stations.

'That's an entire platoon wounded and killed in a single day,' I realized.

Things got so bad, that my platoon was put on standby to head out there. When these orders came in, I was really scared – the most scared that I have ever been in my entire life and the most scared that I ever was during combat, even including the times that I was

114

shot at or near vehicles or areas that were bombed or attacked with indirect fire, such as mortars or rockets.

Why was I scared? In Abu Ghraib, we were supposed to be taking care of the worst part of Baghdad. Now, we were being moved out to Sadr City. More bad than the worst part of Baghdad.

'How bad is "bad" out there?' I wondered to myself.

My platoon and I got our gear, got our vehicles ready, and prepared to move out to Sadr City.

As we were getting ready, I spent a lot of time talking with my gunner. We sat next to each other in the turret during all the times we were rolling in the Bradley. He also took over the Bradley whenever I got out of the vehicle in order to lead Soldiers on dismounted patrols.

About my gunner...he reminded me of my brothers. Actually, he was like a brother to me. He grew up in Kansas City, and he had a young wife and daughter he usually talked to me about. He loved them both.

I used to talk about show business with him. Usually during the down times and when I got bored, I thought about my life as an agent. I would tell him that I enjoyed agenting and that I missed it at times. He understood, so we got along very well.

My gunner said his wife helped to keep him on the straight and narrow and his little girl meant the world to him. He was an excellent Soldier. He knew the Bradley vehicles inside and out, which is why he was my gunner.

My gunner made a call back home. I did the same.

April 5, 2004 (Holy Monday)

We received orders to stand down on the mission to reinforce Sadr City. For the moment, it was a sigh of relief. Our new orders sent us back on our "normal" patrols in and around Abu Ghraib.

April 7, 2004 (Holy Wednesday)

We went out with most of the task force and executed a raid in the wee small hours of the morning, well before sunrise. The primary target was a high-value person. He was not captured, but other known and suspected enemies were captured.

Following the raid, my platoon assumed control of the patrol base in our sector. We were to secure the patrol base and execute patrols for the next 24 hours. We arrived at about eight in the morning.

After lunch, Woody took out a patrol in Bradley vehicles. Shortly after they departed, the marketplace in the area began to empty. This was very unusual, so I radioed Woody. He acknowledged and advised that all was well with the patrol. They were out for about an hour and then they started to head back. As they were getting ready to come down the last stretch of road before turning into the patrol base, they came under small arms fire and were attacked with rocket-propelled grenades.

"Red One, this is Red Four, contact, over," Woody radioed.

"Red Four, this is Red One, copy, what kind of contact? Over," I replied.

"Red One, this is Red Four, taking RPG and small arms fire, over," Woody said.

"Red Four, this is Red One, roger, out."

Woody and his wingman returned fire and neutralized the threat. It seemed that the shooting was over, and they returned to the patrol base. No one was injured, but there was some minor damage to the vehicles.

"It sure is nice having a Bradley," Woody told me, as he pointed to where the vehicle had taken shots.

"Yes, it is," I replied.

I remembered what my counterpart told me when I first got into the country: "We have the armor. We have the firepower. We can handle this."

I ordered my platoon to send the Iraqi Army, Iraqi Police, and all the civilians that were in the compound where our patrol base was, to leave. Our patrol base was also the "city hall" for the neighborhood. I did not want the civilians to be harmed and our location was definitely an enemy target. I was expecting the situation to get worse.

The Iraqi Army, Iraqi Police, and the civilians left quickly and did not resist our orders. At this time, the Iraqi Army and Iraqi Police were relatively new and very inexperienced. I strongly felt that keeping them there at that point in time would cause more problems than it solved.

For a little while, it was quiet.

At about three-thirty in the afternoon that changed. Thump-Thump-Thump-Thump-Thump-Thump. We heard mortar rounds being launched. I was sitting at a machine gun position with some of my Soldiers, and we saw some smoke clouds too. I picked up the radio, but as I got ready to key the microphone, I heard the following, "Six mortar rounds just impacted."

The mortar rounds were landing at a logistics base in our sector.

I keyed the mic and said, "Thunder Mike, this is Annihilator Red One, we observed six mortar rounds fired, over."

Task Force Headquarters acknowledged my report and requested further information. I provided the locations that we observed the rounds fired from and requested artillery fire. My requests for indirect fire were rejected at a headquarters above the task force level.

Then, our patrol base took mortar rounds.

"Red One, Red One." My handheld radio crackled.

"Red One, we have a whiskey India alpha (wounded in action)," I was told.

This was my second-to-worst nightmare. My worst nightmare was to have one of my Soldiers killed in action.

"Go ahead with the Battle Roster Number, over," I replied.

I received the Battle Roster Number, but then I was told that the Soldier was Returned to Duty.

It was a great relief; I acknowledged the news.

A couple weeks later, I learned that one of the Iraqi Interpreters was also wounded from mortar fire that day. Rather than request medical attention, he simply went home with shrapnel in his belly.

"I knew you were busy," he later told me.

We were still under attack, with both mortar fire and small arms fire coming at us. The logistics base that received mortar fire earlier was still receiving it as well.

I called task force and requested artillery fire. My call for fire was rejected. Again, this occurred at a level above the task force.

The fighting raged on. Several of my Soldiers, including one we called Scarface, Woody's gunner, and my gunner all destroyed and neutralized multiple enemy teams armed with small arms and other weapons, such as rocket-propelled grenades. As we kept repelling enemy attacks on the patrol base, we continued to be on the receiving end of indirect fire in the form of mortar rounds. The logistics base was also taking mortar rounds and with greater frequency and quantity.

I radioed another fire mission, but it was rejected.

Task Force Headquarters did send out a platoon of attachments from another unit. I had them pull security on the western boundary of our sector overlooking a key bridge on one of the main roads. Task Force also got a team of Kiowa helicopters to fly for us for a little while. Unfortunately, the helicopters were being shot at when they flew in low to get visual identification of the enemy. I told the helicopters to raise their altitude. The last thing we needed was for the helicopters to be shot down. Due to the fact that the helicopters could not visually identify enemy positions, we were not able to get them to fire on any targets, but at least the presence of the helicopters bolstered our spirits.

My Soldiers continued to repel enemy attacks on the patrol base utilizing both the main guns on the Bradleys and the machine guns within our perimeter.

As day turned into night, we were still engaged in combat with the enemy. My company commander arrived at the patrol base. I was tasked with leading a reconnaissance into the dairy plant across the street. We already knew that the enemy was present there, but we were trying to determine the exact location of the enemy mortar firing positions.

I took two Bradleys and some of my dismounts on the mission. We took small arms fire both on the way in and on the way out, but thankfully, no one was injured, and the Bradleys were not damaged. We did not find any enemy mortar-firing points either.

In other parts of the sector, the task force was taking casualties during the battle.

April 8, 2004 (Holy Thursday)

After about two or three in the morning, things got quiet. The company commander was called back to the Task Force

Headquarters for a morning meeting. It seemed that there would be some quiet.

Third Platoon arrived to relieve my platoon of the security mission at the patrol base. My platoon would move to the logistics base, pull security, and be prepared to execute reinforcement or a quick reaction force mission.

As my platoon pulled out of the patrol base, all hell broke loose. Small arms fire, rocket-propelled grenades, and indirect fire were coming in from three directions at the patrol base. My platoon was outside the patrol base. We were fortunate, because we were mounted in the Bradleys. We were also unfortunate, because we were in between Third Platoon's position, which was under heavy fire and two of the three enemy positions.

As we passed through the fire, I looked back on another Soldier we dubbed Sniper, who was commanding one of the other Bradleys in my platoon, which had some serious mechanical problems. He just looked at me, laughed, and waved. We kept the platoon moving.

As my platoon was about to pull into the logistics base, I got a call on the radio from the company commander. "Red One, this is Annihilator Six, I need you to recover wounded personnel at the patrol base, over."

"Annihilator Six, this is Red One, roger, over," I replied.

"Red One, this is Annihilator Six, report when complete, out," the company commander radioed.

I ordered Sniper to take the Bradley with serious mechanical problems into the logistics base. There was no point in bringing that vehicle back through the fire.

The rest of my platoon turned around and went back to the patrol base. While we were waiting outside the patrol base for Third Platoon to load their wounded Soldiers onto a vehicle, my platoon fired nearly everything we had left at the enemy. We came dangerously close to running out of ammunition for the main guns on the Bradleys during that timeframe. I also came close to running out of ammo for my M4 rifle as well.

A young Soldier in Third Platoon organized their casualty evacuation. He then commanded a vehicle out of the gate to link up with my platoon. He did all of this while under heavy fire.

As soon as he got the wounded loaded onto a vehicle, we proceeded from the patrol base to the logistics base.

At the logistics base, my platoon assumed defensive positions to pull security on the northwest flank of the base. The Third Platoon Soldiers went straight to the field hospital. There was even more small arms fire, rocket-propelled grenades, and mortar fire from the enemy on this day.

The task force was sending almost all of its units into the fight that day.

Apache and Kiowa helicopters were flying support missions and they were being shot at. One helicopter took too many shots and landed at the logistics base. Another helicopter was shot down and it landed just outside the patrol base.

From the logistics base, my marksman, El Toro, and Sniper identified enemy spotters on the roof of a building 1,200 meters away. Sniper asked if they could engage the enemy. I responded, "Go ahead, this is a war."

The first shot missed.

Sniper helped El Toro make the adjustments. El Toro's second shot hit. The enemy spotter was dead. Due to the array of friendly forces, I could not use the main guns on the Bradleys from our position at the logistics base, due to running the risk of shooting other members of our task force that were in the fight. In spite of that, my platoon used machine guns and our rifles to suppress the enemy.

In the afternoon, we were told to return to the forward operating base in order to refit and refuel.

April 9, 2004 (Good Friday)

As my platoon was getting ready to return to sector in the late morning, we saw a massive cloud of black smoke coming in from our area of operations. The dark cloud covered much of the forward operating base. As we were heading out to the patrol base, we learned that a supply convoy was ambushed and was in very bad shape. Second Platoon was already at the site of the ambush recovering killed and wounded personnel.

Fighting in the sector was still ongoing, but the situation was improving. The use of artillery was finally permitted by higher headquarters. The enemy mortars were silenced. This was followed by tank patrols taking out key enemy positions. Our task force was regaining the initiative.

Unfortunately, the initial reports about the supply convoy were not good. Several friendly personnel were killed, many more were wounded, and as many as sixteen were captured. It was not until March 2008 that all of the friendly personnel were accounted for. It was then that Staff Sergeant Matt Maupin's remains were recovered in Baghdad. Until that point, the United States Army listed him as "Missing in Action." He was the last member of the convoy to be sent home.

After the casualties from the supply convoy were evacuated and the ambush site secured, full-fledged fighting continued. As darkness fell, our task force regained controlled of the sector.

April 10, 2004 (Holy Saturday)

In the early morning hours, the guns of war went silent.

After dawn, I took out a patrol to escort engineers as they cleared up the debris from the past three days of battle.

April 11, 2004 (Easter)

On Easter Sunday, my platoon spent some time back at the forward operating base. I got a chance to call back home. There was a phone bank set up, which consisted of a trailer with a bunch of pay phones. I called my family as they were having Easter dinner.

They asked, "How are things going?"

I replied, "We've been pretty busy, but all is fairly well."

My platoon went back out to the patrol base before nightfall. We executed a raid and captured an insurgent leader and his two sons. They were all former members of the Special Republican Guard, one of the elite units during the Saddam Hussein era.

We brought all three prisoners back to the forward operating base the following morning.

It was a good way to end the very long week.

The Car Bomb

30 September 2004

I was now serving in the Civil Affairs section at the Division Headquarters. I worked on tribal and religious affairs throughout Baghdad. I also led a small team that escorted high-ranking officers to meetings with sheiks and clerics.

On the morning of September 30th in 2004, I was drinking coffee when I felt the ground shake and heard a loud boom following. In Baghdad, that meant only one thing – a vehicle-borne improvised explosive device or VBIED (a car bomb). I went to the Operations Center and listened as the radio reports came in.

As I heard the reports, my heart sank. I realized that the patrol base that I used to operate out of when I was a platoon leader was the target of the car bomb. There were wounded Soldiers and one Soldier had been killed. I knew that it was definitely men from my old company, but I wondered if it was my old platoon.

Later in the afternoon, I learned that Soldiers from my old platoon were casualties from the car bomb attack.

Fortunately, most of the wounded Soldiers were returned to duty, but one of the Soldiers was severely wounded and at the hospital. I got on a convoy and went out to the hospital.

My old Soldier proceeded to explain to me that his ex-wife was spending all his money and that he needed help from a lawyer. "Can you talk to JAG?" He asked.

"Of course," I said, "You got it."

The first stop I made when I got back to headquarters was to the lawyers. I explained the situation to them and asked them to do whatever they could to help. I also told the lawyers that my old Soldier was only 20 years-old and that he was going to lose the use of one of his arms.

The even harder news was finding out who was killed. The Soldier that died was Specialist Rodney Jones and he was known as "RJ." RJ wanted to be the Mayor of Philadelphia after he finished his time in the Army. He was intelligent, patient, and adaptive. He exemplified the counter-insurgent Soldier of today's Army. RJ learned Arabic from the Iraqi Soldiers that worked with our platoon. In fact, RJ got so good at speaking Arabic that he could tell me whether the interpreters were doing their job correctly or not. He got along well with everyone. He was quick to pick up new assignments. On his last day, he was standing guard at the main gate to the "city hall" for Abu Ghraib. A truck with over 800 pounds of explosives rammed into the Bradley Fighting Vehicle that RJ was in. The explosion from the car bomb killed him instantly. RJ was trying to keep one part of the young Iraqi government intact.

For me, this was the worst day of the war.

The Hostage

October – November 2004

In October of 2004, shortly before the major military operations in Fallujah, members of Al Qaeda in Iraq captured the brother of one of our local advisors in Al Anbar Province. Due to the fact that we had taken care of his brother, the hostage had intervened when members of the Coalition Forces were taken hostage earlier. Apparently, his intervention was noticed and it pissed off Al Qaeda in Iraq. Therefore, members of Al Qaeda in Iraq hired a crew to conduct the kidnapping.

The hostage was well regarded in Al Anbar. At the time, the Sheiks in Al Anbar were willing to vouch for his release, if we provided military assistance to them. The Sheiks feared that without American support, they alone were not strong enough to fight Abu Musab Al Zarqawi and his nascent Al Qaeda in Iraq organization. The Sheiks were also concerned about being left out of the political process, because they had boycotted the ballot initiatives supporting the upcoming elections.

We requested support for the Sheiks, but our request was denied.

In the fall of 2004, the military leadership in Baghdad did not have much trust in the tribal leaders of Al Anbar Province.

Then, the old school Green Beret, our sergeant, and I requested to execute a hostage-rescue operation in Fallujah. This request was also denied.

In November 2004, the Marines led a major offensive into Fallujah. The hostage was recovered during these operations. Initially, he was imprisoned at one of the Coalition's detention facilities. A couple weeks later, the hostage's name showed up on a detainee roster. Eventually, we were able to facilitate his release.

It was a happy ending, but an extremely frustrating chain of events. Two years later, many of the same sheiks that were willing to vouch for the hostage, served as the core leadership of the Al Anbar Awakening during "the Surge."

The Surge was the time in Operation Iraqi Freedom when the number of troops increased. In addition, the Soldiers were moved from large bases to smaller camps in order to be better integrated with the local populace. Much of this was executed as part of General David Petreaus's strategy of counterinsurgency in Iraq.

The Christian

December 2004

In the Civil Affairs section, my boss was an old school Green Beret. He was very much out of the mold of blending in with the locals in order to help them win their fight. He was also very sympathetic to the locals. This old school Green Beret was a Special Forces Soldier that firmly believed that the core of Special Forces was working foreign internal defense and unconventional warfare. He was a believer in counter-insurgency (COIN) long before it was brought into the Army mainstream by General David Petreaus.

In December 2004, an Iraqi Christian woman that was working for our Civil Affairs staff was being followed while going to and from work. At the time, many translators and other local staff supporting US and Coalition Forces were being followed and killed.

The old school Green Beret feared for the life of this woman. Since she was a Christian, she stood out like a sore thumb. Christians are a very small minority in Iraq. There was also an increase in violence against Christians in Iraq at this time.

Everything had been so real and so intense thus far that I had not had time to reflect too much about my time as a Hollywood agent.

But knowing that I had worked in the film industry, the old school Green Beret asked me to find the Christian a job working in the movie business.

"You used to work in Hollywood," he said, "so find her a job."

"It's not that easy," I replied, "but I will try."

"She looks like Sandra Bullock," the old school Green Beret insisted.

"A little bit," I *tried* to agree.

I made some calls and sent some e-mails. There was a chance that I could get this Iraqi woman into the United States. I contacted my mentor, the one I used to visit on the Paramount Pictures lot. He was open to trying to line up a job for her.

I informed the old school Green Beret of this possibility.

"Wait a minute. If she works in movies, then will men try hitting on her?" The old school Green Beret asked.

"Most likely," I replied.

"Then scratch that plan," he instructed.

I called my contacts and told them all bets were off.

"Why don't you marry her?" The old school Green Beret asked.

"Marry her?" I replied.

"Yes, that way she can get out of here," the old school Green Beret said.

Well, I did not marry her and as far as I know, she never went to the United States. And she also survived her job working for us. Not too long after the new year, she quit her job. It was understandable. The risk of working for us out-weighed the financial rewards.

The Little Girl

January 2005

In December of 2004, I went on my first trip to see a little girl. The goal of our mission was to assess the little girl's medical care. The little girl was five years-old and she had bullet and shrapnel wounds in both of her legs. She was a victim of being in the wrong place at the wrong time. In the same firefight that she was wounded, the little girl's sister was killed.

The old school Green Beret actually met the little girl at one of our bases shortly after the firefight. She received her initial medical care from American medics. The medics told the old school Green Beret that, provided the proper medical care, the little girl would keep both of her legs. However, the policy governing medical care limited the amount of care that US Forces could provide to non-Americans. Therefore, the little girl's care was placed into Iraqi hands. The good news is that there are several well-trained doctors, nurses, and medical staff in Baghdad. The bad news is that the Baghdadi hospitals were woefully under supplied with basic items, such as cleansers, antibiotics, and bed linen. Due to this, there was a high likelihood that the little girl's legs would remain infected and eventually require amputation.

As we walked into the hospital where the little girl was being treated, the first thing that struck me was the smell of decaying flesh. The second thing that struck me was that almost all of the patients were trauma victims: Iraqi Soldiers, Iraqi police, and civilians that had been shot, blown up, or something in between. The little girl's doctors were compassionate, but they did not have what they needed.

We returned to base and wrote a report about her status. The critical question was whether she could receive more care from Americans. We endeavored to make that happen, and we succeeded, at least for a follow up.

The red tape was cut and we were allowed to take the little girl to see the same medics and doctors that initially treated her. After visiting them, she was much better. So too was her mother.

"She needs American care...back in the States," the doctors told the old school Green Beret.

How do we do that?

"We have to get her back to the United States," the old school Green Beret declared. "Link up with the lawyers and see if we can get her family paid for her injuries and the death of her sister."

Shortly after the start of the war in Iraq, the military started paying solatia (payments that are made to right a wrong to the family of someone). Solatia payments were made to families of victims that were inadvertently wounded or killed by US Forces. This was consistent with local customs. In order to make the payments happen, an Iraqi had to request them and an American military lawyer had to review them. The bottom-line was that someone had to be hurt and ask for help and the financial payment had to comply with several policies.

I met up with some lawyers that I knew. They were great guys.

"How can we help?" They asked.

"Here's the deal…" I went on to explain the situation about the little girl.

"First, you need to get her family to ask for the payments. That part should be easy," the lawyers told me.

"Fair enough," I replied.

"Then, we need to make sure that there were no Rules of Engagement issues. This probably is not a big deal. We also we need to show that the family was shot accidentally, and that might be tricky," they said.

"Explain, please," I asked.

"The deal is this, if she was caught in the cross-fire during a gun-battle, there is no payment. If she was shot by stray rounds or shot intentionally, such as mistaken for bad guys, there is a payment."

"I see."

I went back and advised the old school Green Beret. We went out to see the little girl's mother and she requested the payments. We got it in writing.

I took the mother's request back to the lawyers.

"Now, we can get the ball rolling," the lawyers told me.

"How long do you think this will take?" I asked.

"Probably a couple of weeks," they said.

With that, we were on the way to getting the little girl's family some money to pay for their suffering.

"We still need to figure out how to get her back to the United States," the old school Green Beret said.

Through some serious cajoling, we were able to get permission for the little girl and her mother to fly back to the United States in order for the girl to get her medical treatment. The only condition was that the Public Affairs staff needed to be aware of the story. Ultimately, everyone agreed. The little girl and her mother allowed the military Public Affairs personnel to take notes and pictures.

The lawyers advised that the payments were approved. The maximum payments for both the death and the injury were authorized. We brought the little girl's mother back to our base to meet the lawyers that arranged the deal. Without a doubt, she was incredibly grateful. I was happy we could do something for them.

The little girl went to the United States, got the proper care, and her legs were saved.

The First Iraqi Elections

January 2005

All of the activities surrounding the "get out the vote" were being led by the United Nations. On the American side, this meant State Department personnel. The security effort was going to be a joint operation between the Coalition and the Iraqis. It was really the first major joint operation between Coalition and Iraqi forces.

To be honest, I was highly worried about the elections. Due to all the security problems, I expected that many of the Iraqis would be too scared to vote. In addition, most of the Sunni population was postured to boycott the elections. In spite of the fact that some Sunni leaders "saw the light" late, the Sunnis failed to get any of their political groups placed on the ballot on time.

One day, I was tasked to ride over to the Baghdad International Airport with one of my co-workers. Our mission was to confirm that that the United Nations's security firm was on site with the ballots that had just landed. My co-worker and I jumped into an SUV and rode straight across the runway at top-speed. Planes were taking off and landing at the time. This was a chance to live out one of those, 'You will never guess what I did, this one time...' war stories.

Then, we conducted the link-up with the security firm and the ballots were already loaded and ready to roll to the polling places. My co-worker and I returned to base and got to tell the story of driving across the airstrip like a bat out of hell.

When Election Day came on January 30th of 2005, it was amazing. People walked miles to vote and in many places, literally risked their lives. The Iraqi police and military did a superb job, along with the Coalition, in securing areas surrounding polling stations. 58% of the eligible voters cast their ballots to form the body that would be tasked with writing Iraq's Constitution.

The elections were a success.

Chapter 6: Coming Home

"I told you to wait for my signal, you didn't wait for my signal." –

Jane

"Well, I improvised." – John

From: "Mr. and Mrs. Smith"

December 2004 – November 2006

The Lieutenants

December 2004

Before the first Iraqi elections, I had been assigned a partner. Her name was Dawn.

Dawn was another American Soldier who became my work partner. Dawn was from Pennsylvania originally and graduated from high school in North Carolina. She enlisted in the Army out of high school. She served in Honduras, Panama, and the first Gulf War during her first few years in the Army. Later, she served in Korea, went to the Officer Candidate School, and arrived in Iraq as a staff officer in a Civil Affairs battalion.

As far as our initial "meeting," it was somewhat awkward. I was told that she was taking over a boring job that was one of the three jobs I did. However, Dawn was told that she was taking over one of my jobs, but that I was keeping the one boring job.

Dawn was sharp, fierce, and beautiful. She had blue eyes and dark hair. I did immediately notice she was beautiful (and others observed this).

The gist of our first conversation was:

Me: "Hello, I'm George."

I extended my hand.

Dawn: "I'm Dawn."

Dawn reluctantly took my hand.

Me: "Nice meeting you."

We shook hands.

Dawn, letting my hand go said, "You too."

Me: "Let me know if you need anything."

Later, I learned that Dawn initially found me to be a snob, because I carried a leather-bound notebook, which was a gift from one of my "show business" friends, Marc.

Eventually, Dawn and I became very effective partners and good friends. We developed an excellent working relationship and a logical division of labor. We analyzed tribal and religious relationships. We also picked up an additional task – cultural awareness training.

There were two parts to the cultural awareness training. One element of the training was to increase awareness amongst the Soldiers already in Baghdad. An example of this is describing how the locals sleep on their roofs in the summer in order to be cool. The second element was to develop awareness in the Soldiers that would soon arrive in Baghdad. An example of this was describing the demographics of the areas that the Soldiers would patrol.

In order to train the Soldiers that would arrive in Baghdad, a team was assembled to travel back to the United States. The team consisted of three Iraqi cultural awareness trainers and a Civil Affairs Soldier escort. For reasons not known to me, the method of travel was commercial rather than military airplane out of Baghdad.

A day before the team was set to leave I drove them over to Baghdad International Airport in order to get their tickets. Upon entering the airport area, the security guards required me to clear my weapon – remove all ammunition from it. I put my magazines into my ammo pouches on my vest. The security guards also directed the Civil Affairs Soldier acting as the escort to leave his weapon with me. This was not a good sign.

I then parked my vehicle in the garage and waited for the team to get their tickets. This was a bad choice on my part. I had an unloaded weapon and four armed Iraqis, all within fifty meters of my vehicle, surrounding me. As time ticked by, I was getting more and more anxious.

After fifteen minutes, I thought, 'How long does it take to buy tickets?'

After thirty minutes, I thought, 'When are they coming out? We should have brought the entire security team on this.'

Then, one of the armed Iraqis approached my vehicle. He came up to me. Another armed Iraqi approached the vehicle on the passenger side.

At this moment, I was regretting the fact that my M-16 was not loaded.

A quick glimpse in my rearview mirror and I confirmed there were two additional armed Iraqis with locked and loaded AK-47s.

In broken English, the guy that approached the driver's side started a conversation with me.

Officially, this guy was airport security, but he was not even reminiscent of a rent-a-cop. Far from it. He approached me.

Through his broken English and my broken Arabic, we muttered through thirty minutes of a conversation. We even talked religion and politics – very taboo, but there were no problems.

Then, he asked me for money.

"I can get you girls," the airport security guard told me.

"La, shukram," I replied. Translation: "No, thank you."

"I can get you beer, whiskey, drinks," he told me.

"La, shukram," I replied again.

"I want money. Ten dollars," he told me.

"I can't do that," I told him.

Out of the corner of my eye, I saw the people that I brought to the airport.

"Masalama," I said. Translation: "Goodbye."

The guards backed off and pulled out fast. I picked up the team and we headed back to the base.

"We have to do this a little differently next time," I told my boss.

"For starters, we will keep our weapons locked and loaded and there will be two vehicles," I said.

My Civil Affairs boss agreed.

I informed everyone what had happened, and what could've happened. Then worked out the rest of the details with the various people that were part of the mission to make certain next time, we were all better protected.

The Mission

December 2004

After the ticket debacle, we had another setback. The commercial airline personnel would not let the Civil Affairs Soldier acting as the escort, board the plane. Our escort did not have a passport, only his military identification card. Because of this, he was not allowed to board the plane. Meanwhile, the three Iraqi cultural awareness trainers got on the plane and left Baghdad. Their first stop was in Amman, Jordan.

We did not find this out until our escort showed back up at our base.

"Where are the Iraqi trainers?" The Old School Green Beret asked.

"On the plane to Amman."

Immediately, the Old School Green Beret asked for any and all people that brought their passport to step forward. Very few of us had brought our passports, but Dawn did.

Most people do not travel with a passport on deployment. Generally, you go the combat zone and then you go home.

"Dawn, you will link up with the team in Amman, and I will go with you," the Old School Green Beret said.

While Dawn was getting things ready, I grabbed her some food from the chow hall. Later, she told me that this was the initial icebreaker for her with me. I had thought nothing of it. She had a mission and she needed food, so I got her food while she got ready.

The security team and I took Dawn and the Old School Green Beret to the airport and ensured that they got on the plane. They landed in Amman, Jordan without incident and linked up with the three Iraqi cultural trainers. For the most part, the first trip went alright. The, the old school Green Beret returned to Baghdad. This left Dawn alone to handle the mission with the Iraqi instructors for the duration of their time in the United States.

Christmas in the Middle East

December 25, 2004

Dawn was returning from the United States on Christmas. She ended up spending Christmas in Amman, Jordan with the Iraqi cultural awareness instructors that she was escorting. The cultural awareness instructors had enjoyed their trip in the States. They wanted to thank Dawn, so they did their best to "guide" her in Amman during the layover.

While in Amman, Dawn received a complimentary breakfast from the hotel. At the breakfast, the hotel staff was selling time shares in Beirut. Dawn's thought about all this was, "I guess they do time shares the same way everywhere."

Meanwhile, I led a security team for the Old School Green Beret to Baghdad's International Zone which was used by coalition forces and civilian authorities and subject to high security, also referred to as the Green Zone. The Old School Green Beret was meeting with some leaders to discuss ways to reduce the sectarian tensions. We had Christmas dinner at the Al Rashid hotel and returned to base.

We, the American Soldiers, were lavished with packages from the United States. This included gifts from our families and friends, but also gifts from strangers that just wanted to say that they cared.

On Christmas, we were busy, but everyone got a chance to celebrate or reflect in some way.

A New Year in Baghdad

January 1, 2005

Dawn returned after Christmas, but before New Year's Day. We did a lot of work together and had become better, closer friends. Left unsaid, was the fact that we were developing an attraction to each other. I found Dawn attractive the moment that I first set eyes on her. I repressed my feelings, because she was a co-worker. Dawn told me later that she started to like me after I got her the meal from the chow hall.

In spite of all this, we were very busy with the tasks at hand.

Due to the increasing sectarian tensions, we became much busier with regards to both tribal and religious analysis. Determining who the key leaders were and trying to get them to talk with each other rather than kill each other was the primary objective.

A great many of the tribal leaders feared both Tawhid Al Jihad, which became Al Qaeda in Iraq (led by Abu Musab Al Zarqawi at the time) and the Iranians. The primary focus of the tribal leaders was the safety and security of their people. The religious leaders were also interested in peace, but many of them were seeking a stronger position in the fledgling Iraqi government.

On New Year's Day in 2005, Dawn and I met with one of the tribal leaders, a Sheik, at his Baghdad residence. The Sheik, like many of his fellow tribal leaders, was concerned about both Tawhid Al Jihad and Iran. We were meeting with him in order to capture his concerns and determine who from either the military or other agencies of the United States Government might be able to follow-up with the Sheik.

While Dawn had met the Sheik before, this was my first time meeting him. I wanted to make a good first impression. The Sheik offered me a cigarette. I don't smoke, but I had a lighter that was a mock-up of Indiana Jones's from *Indiana Jones and the Last Crusade*. Out of respect, I accepted the Sheik's offer. Then, I pulled my lighter out to light the cigarettes. I literally choked the cigarette down. The Sheik, politely, did not ask if I wanted another one.

After the meeting, Dawn told me that she was amused with my approach.

The meeting itself went well. Unfortunately, engagement with the tribes was not a high priority for our efforts at the time. Dawn and I wrote our report and sent it up the chain-of-command, but we were not able to determine who could follow up with the Sheik.

Under the premise of 'something is better than nothing,' Dawn and I maintained contact with the tribal leader in order to keep open the line of communication and potentially set conditions for greater engagement and cooperation at a later time.

The Mission Continues

January 2005

Our cultural awareness-training mission for American Soldiers also continued. A couple of weeks after the New Year, a second trip was planned for the Iraqi cultural awareness team. They would travel to the United States in order to train the next unit of soldiers that were coming into Baghdad.

Dawn would serve as the military escort for the three Iraqis in the cultural awareness team. I insisted that another military officer travel with Dawn in order to escort the team. In the Army, we have the buddy system. Part of the rationale is that your buddy watches your back and vice versa. I was concerned about Dawn's safety en route to the United States. My logic prevailed. Another officer from a different staff section was identified as having his passport on hand. He and Dawn were now the escorts for the mission.

Once again, the security team and I took them to the airport.

This time, things were different.

Right before she left, Dawn told me, "Don't get too deep about this, but I will miss you."

I told her that I would miss her too.

155

Then, the smart-ass in me asked, "So, what do you mean – too deep?"

She just laughed.

The security team and I ensured that Dawn and the cultural awareness team got on the plane to Amman again. After they boarded, the security team and I returned to our vehicles and rode back to base.

My head was spinning.

'I think she likes me,' I thought.

The Problem

January 2005

Upon landing in Amman, Jordan, two of the three Iraqis were arrested by the Jordanian authorities. The reason given was that they were both Shiites and Jordan reportedly was not allowing Shiites to enter the country.

Dawn called me to report the situation.

"See if you can get them back here. Then, you get back here," I told her.

I told my superiors that two of the three Iraqis were detained in Amman, Jordan. Then, I told my superiors that I had instructed Dawn to have everyone return to Baghdad as soon as possible.

Surprise is often part of war. Though you always want to have the element of surprise on your side. In this situation, surprise was against us. Furthermore, Jordan was not part of the conflict. So, there were diplomatic and political considerations regarding the options that we had to get everyone back to Baghdad.

There was silence for quite some time (nearly a minute), then my superiors started going through a list of things to do. Most of these things, I had already done. We called the US Embassy in Baghdad. We called the US Embassy in Amman, and we called some other military and government officials in the region. In spite of all this, there were very few real options to get everyone back quickly.

Diplomacy does not work fast. Going in with force was not approved. The best option was to obtain confirmation that the Jordanians would allow the Iraqis to fly back (as Dawn arranged) and then confirm and re-confirm that Dawn and the other American officer were safe.

One of the main questions asked during the course of all the conversations and meetings was, "Are the Americans all right?" Dawn had been emphatic that she and the other American officer were safe.

The next day, the two detained Iraqi instructors returned to Baghdad. I picked them up at the airport. The Jordanians kept their promise. Both of the Iraqis were shaken up, but they were not beaten or harmed in any way. I called Dawn to tell her, and she was relieved.

Following this, Dawn and I had several phone conversations. She was still in Amman with the other American officer and the third Iraqi instructor. The three of them were scheduled to return the following day.

Over the course of talking on the phone, Dawn and I reviewed the situation. We knew that she and the other two returning to Baghdad would resolve the situation.

But as we talked, in that moment, I knew we had feelings for each other. I told Dawn that I liked her and that if we were not in the middle of a combat zone that I would ask her to go on a date. She told me that she would want to go on a date too.

The following morning, Dawn, the other American Army officer, and the third Iraqi cultural awareness instructor returned to Baghdad without incident.

Later, the security team and I escorted the Iraqi to his home in the "red zone" meaning that it was in an unsecure area, where the enemy might attack (either us or him). Then, we took the other officer to his quarters.

After that, I walked Dawn back to her quarters. At this point, we both acknowledged that we had strong feelings for one another.

But even though we now knew that we wanted to be together, we still had a mission to accomplish. Training and experience told us both that we could not lose focus on the mission; this was especially true as violence in and around Baghdad increased.

The Olive Branch

January 2005

On one occasion, Dawn, the Old School Green Beret, and I
conducted a meeting with an intermediary. This intermediary was
a person that several tribal leaders had met with before telling us
about him. The intermediary was reportedly authorized by some
of the insurgent groups fighting U.S. forces to meet with the
Americans. How he was connected to the insurgent groups was
not clear.

Dawn, the Old School Green Beret, and I met with this
intermediary at a neutral location. The insurgent groups
dispatched him to say that they were ready to stand-down, if the
United States guaranteed the Sunnis a place in the political
process.

"We cannot do that. The Independent Electoral Commission is
responsible for the elections. However, if the people that sent you
do stand down, then we can outline how they can get involved
prior to the next elections, participate with the soon-to-be-elected
government, and work out something with both the military and
the our State Department to ensure that the leaders are not captured
by Coalition Forces." It was a lengthy reply, but it was the best
that we could muster.

"This is not what the people I represent wanted to hear, but perhaps they will consider this. We must keep these lines of communication open," the intermediary said.

At this point in time, there were several issues facing the Sunni Arabs in Iraq. The first issue that they faced was the impending lack of representation in the future government. This issue was rooted in their boycott of the first elections. By the time the Sunni Arabs in Iraq realized that participating in the electoral process was worth their while, it was too late, because they had missed the filing deadlines and were essentially out of the process. Since the Department of State had the lead, we (in the military) were very limited in affecting this issue.

The second issue facing the Arab Sunnis was directly related to the first issue. By being outside of the political process, the Arab Sunnis (a demographic minority) would soon be subordinate to the Arab Shiites. While some of the Sunnis and Shiites shared bonds of kinship across tribal lines, it was not enough to overcome the larger concern – the influence of Iran in the future or Iraq. This issue was both international and primarily political in nature. Again, the Department of State had the lead and the military's options were limited.

The third issue facing the Arab Sunnis was the most immediate and problematic. Tawhid Al Jihad exploited and terrorized the Arab

Sunnis. As explained earlier, Abu Musab Al Zarqawi led this group and they eventually became Al Qaeda in Iraq (AQI) and later the Islamic State in Iraq and Syria (ISIS). Tawhid Al Jihad was kidnapping, raping, and killing people in the Sunni Triangle of Iraq. The Al Anbar Awakening and the Surge eventually resolved this issue. However, in early 2005, this was a big problem. The Arab Sunnis were seeking American commitment. Initially, the American support was limited, but as described earlier, this situation turned around during the Surge. Unfortunately, this situation deteriorated after the withdrawal of US Forces from Iraq and that was evidenced by the presence of ISIS extending from Raqqa, Syria to Mosul, Iraq.

At the time of the meeting with the intermediary, there were limited options from the US military regarding the key issues presented. While nothing was resolved, this meeting served as an additional indicator of the status of the Arab Sunnis in Iraq. Furthermore, I believe that it shaped conditions for the future, if only to provide critical information to decision makers many months later.

Similar to the meeting on New Year's Day with the Sheik, the meeting with the intermediary served as a potential stage setter for future engagement by the military or other agencies of the United States Government.

We completed the meeting. The intermediary and I got along well. For whatever reason, he seemed to have a certain level of trust in me. In Iraq, it is customary for men to hold hands. After everyone said goodbye at the meeting, I escorted the intermediary out of the secure area, walking arm-in-arm. The US Soldiers standing guards must have thought that I was a few cards short of a deck. After all, I was walking into the red-zone in plain clothes, without body armor or a weapon, although, I actually had a pistol tucked under my belt, and I was arm-in-arm with an Iraqi guy.

Nothing happened. The intermediary got out safe.

Contemplating Life After Baghdad

January 2005

I should mention though, during that same meeting with the
Intermediary, something did in fact happen, but it was between
Dawn and I...

During a lull in the meeting, Dawn passed me a note.

"Want to get married?" She wrote.

I was completely surprised, but extremely happy.

"Yes," I wrote back. "Let's try to focus on the business at hand
here."

After the meeting with the intermediary, Dawn and I discussed
what had happened, the note...

Me: "So, do you really want to get married?"

Dawn: "Yes."

Me: "Good, me too."

Dawn: "Well, when are you going to propose?"

Me: "After we get back home. I love you."

Dawn: "I love you too."

During dinner that night, we talked seriously about the note, but also joked about the timing and delivery of it. The conversation was also more about when and where we would actually get engaged more than it was about marriage and beyond.

We planned to take a big trip after we returned to the U.S. from Iraq. We were looking forward to getting back to the States.

We discussed "life after Iraq" extensively. Dawn told me about the house she had in the Poconos in Pennsylvania. We started talking about things we could do when we were back in the United States and out of the Army. I talked about going back into show business, in either Los Angeles or New York. Dawn talked about going into government work in either DC or New York. We thought that maybe New York was where we needed to go.

Before I met Dawn, I started putting together an assignment request to work at the Institute for Creative Technologies in Southern California. This would bring me back to L.A. But after meeting Dawn, I stopped working on that paperwork and instead I put in a request to stay in Baghdad until Dawn left. If approved, then she and I would return to the U.S. together.

My request was approved. Now, rather than returning to the U.S. after thirteen months in Baghdad, I would depart several months later.   It was set, Dawn and I would return to the United States together.

Babysitting In Baghdad

February 2005

Dawn and I visited a refugee camp. While there, the senior officers that we were with spoke with the adults, many of them widows, and Dawn and I watched the children. We would play soccer with them, give them candy, and just try to help them feel more comfortable.

Growing up in a war zone must be awful. There were generations of Iraqis that knew nothing but war. In the 1980s, there was the Iran-Iraq War. In the 1990s, there was the Gulf War, followed by uprisings, oppression, and international sanctions. Then, there was Operation Iraqi Freedom. This meant that the Iraq 20-something and younger population came of age to the backdrop of armed conflict.

Some of the people that Dawn and I worked with joked about us being the highest paid babysitters on the planet, but it didn't bother us. All in all, it was part of the mission. We were content on doing it, helping people survive the horrors of the war.

Engaged

June – September 2005

Dawn and I spent our last few months in Iraq working with sheiks and clerics, facilitating some humanitarian assistance, and pondering life back in the United States.

One day, I was browsing the internet for engagement rings. One of the officers that we worked with saw us.

"So, George what are you doing?" He asked.

"Oh, nothing," I replied.

Dawn was sitting caddy-corner to me.

"It looks like you are looking at engagement rings from what I can see," he said.

I started to turn beet red. "Well, that is correct."

"Is the ring for anyone that we know? Anyone that might be sitting here right now?"

Dawn and I had tried very hard to keep our feelings for each other discrete.

After a beat, the officer said, "Look, we all know. Good luck!"

"Thanks," I told him.

Dawn and I returned to the United States in late June of 2005. We planned to take a trip to see the pyramids and see the Nile in Egypt and then spend a week or so in Fujayrah in the United Arab Emirates. Shortly after we booked the trip, terrorists attacked the Red Sea resorts in Egypt, so we cancelled the trip to the Middle East. Instead, we decided we'd venture to the Finger Lakes in upstate New York.

Dawn and I were engaged in late July 2005 in the Poconos. I left the engagement ring on a bed surrounded with rose petals. When Dawn walked into the room and saw it, she got choked up. I got on one knee and asked her to marry me, and she said, "Yes."

A couple days later, we were heading up to the Finger Lakes. On our trip, we decided to take a limousine tour of the wineries in the area. As we were getting close to the end of the tour and very much in need of the ride back to our hotel, the limousine broke down. It was a first for both us. Our driver was a good sport about it. A mechanic came out and got the car in shape to get us back. While it was not the pyramids, we still had a fun and exciting trip.

Hurricane Katrina

September 2005

On Labor Day weekend, Dawn and I planned to have an engagement party at her house in the Poconos in Pennsylvania. We were both very excited to share our news with everyone. Though I had returned from a tour of duty in Iraq, I was still on active duty with the 1st Cavalry Division at Fort Hood, Texas.

As we got closer to Labor Day, however, it appeared that a strong hurricane was bracing to make landfall along the Gulf Coast of the United States. Looking at a map, I realized that several states along the Gulf Coast had deployed most of their National Guard Soldiers to either Iraq or Afghanistan. Implied in that was the likelihood that nearby Army units would have to provide support to the states affected by the hurricane. The closest Army units were at Fort Hood, Texas and Fort Bragg, North Carolina.

When it seemed all but certain that Hurricane Katrina was happening, the Commanding General of the 1st Cavalry Division rescinded all leaves and passes. Drawing the same conclusion that the National Guard had too many units deployed, General Peter Chiarelli told the 1st Cavalry Division, "These are Americans, so we must be ready to help them."

With that, I began getting ready to provide support. Our civil-military operations office would be responsible for identifying and coordinating the interaction with civilian agencies responding to the event. Fortunately, my new boss had just come to Fort Hood from New Orleans, so he was familiar with the area and the civilian agencies. He moved into his new house (in less than a day) and packed his bags to go back to New Orleans.

Because of the severity of the storm, I would not be going to the Poconos for our engagement party that Dawn was now preparing for alone. I missed the party, but Dawn understood and so did all of the guests.

The arrival of the 1st Cavalry Division in New Orleans, Louisiana was big boon to the response and relief efforts. Due to the laws constraining the use of the "regular" Army within the United States, at the time, it took time for the unit to get ordered into Louisiana. After the storm, many lessons learned were identified regarding such constraints.

The Soldiers that went to New Orleans were proud to be part of an effort to assist their fellow Americans directly. In Baghdad, progress ebbed and flowed. In New Orleans, Soldiers were able to see the tangible results of their efforts on the streets.

Amazing Grace

October 2005

While I was in Iraq, my father suffered from cancer. By the time that I was back in the U.S., my father was close to death.

Dawn and I were en route to Rhode Island, on the way to see him in the hospital. This would have been the first meeting between Dawn and my family. I was eager for her to meet my parents. But then I got a call from my brother Matt. My father had died.

I went back to Rhode Island for the funeral.

"No more rough seas…Smooth sailing forever…" the card read. Long-time friends of my family wrote it. The note was attached to the flowers that they had sent for my dad. It made sense in the way that it was simple, to the point, and unending. My dad loved sailing, boats, and the ocean in general. Dad grew up on the water and he and my mother raised my brothers and I on the water too. Mostly due to sailing, boats, beaches, and the ocean, the summer was all of our favorite time of the year. He had also been in the Navy from 1965-1969 and he actually served on the same ship as Dawn's father – the USS Saratoga. Dad was also very honest and he was always available. Dad had succumbed to a painful cancer called mesothelioma.

"Present arms!" The detail commander ordered.

I raised my hand in a salute to my dad beneath the flag draped over his casket.

The Funeral Detail fired the first volley.

The crack of the guns in the salute startled me.

The detail was using either M-1 or M-14 rifles. Both are 7.62 millimeter rather than the 5.56 millimeter M-16 and M-4 rifles that I carried in the Army. AK-47 rifles are also 7.62 millimeter. The sound of the first volley was reminiscent of the AK-47 fire in Baghdad.

That familiar yet uncomfortable sound startled me.

The Funeral Detail began folding the flag in order to hand it to my mom.

The detail commander presented the flag to my mom saying, "On behalf of a grateful Nation."

Once again, salutes were rendered, this time to the flag.

The final respects were paid to Dad, the man who taught me more than anyone else ever did about what it really meant to be a man, a husband, and a father.

During Dad's funeral, I realized that the prayers of family, friends, and neighbors carried not only me, but also, my entire platoon through the fire. For that, I was immensely grateful.

Less than a year later, Dawn's mother Karen would suffer a stroke and pass. It was a shock to Dawn and her family. There had been no indications of even the possibility of a stroke prior to it occurring. I met Karen at her home in North Carolina after Dawn and I returned to the United States from Iraq. She was a kind and generous woman.

Having recently lost my father, I knew the pain that Dawn and her family were going through with the loss of her mother. It was a sad time.

Moving On

January – November 2006

In January of 2006, I left the Army and Dawn and I moved to Alexandria, Virginia. She had gotten a job as a government contractor.

Dawn and I talked about my returning to the entertainment industry. But for me to be close to her or with her, I would have to be near Washington, DC, so that limited my options in film and television from slim to none. I ended up getting a job selling internet services.

We were still re-adjusting to "normal" life.

There were many other Soldiers, Marines, Airmen, and Sailors that went to wars and did not come back. And there were many more still that went to war and came back severly wounded physically, mentally, or both.

It took quite some time for us to reintegrate after we came home, but at least we had each other.

Now, for nearly ten months, I had been out of the Army. But the war in Iraq (and Afghanistan) was only getting more intense. Due to the situation there, the Army was recalling personnel from the Inactive Ready Reserve (which I was now part of).

In the summer, I learned that many of my peers that left the Army shortly before I did, were being called back in. Dawn and I discussed these re-calls. Rather than be recalled to active duty like my friends, we decided it was a better option for me to join the National Guard.

I called a recruiter with the Virginia Army National Guard.

"What do you do?" The recruiter asked.

"I am an Infantry officer," I said.

"Great," he replied, "we have a position in Roanoke."

"Sir," I said, "is there anything closer? Do you have any Civil Affairs or Information Operations positions?"

"Information Operations? As a matter of fact, we have this one unit in Manassas," he said.

The recruiter set me up for an interview with the Commander, and I was accepted into the unit.

The paperwork would take some time, but within a few weeks, I would be part of the Virginia Army National Guard.

After Thanksgiving, our next-door neighbors met us for drinks at a pub in Alexandria, Virginia called Pat Troy's. In the way that a sports bar might have jerseys or photographs of the local teams, Pat Troy's had uniforms from all services of the military, as well as several police and fire departments along its walls. One of the corners of the bar was dedicated to President Reagan and was called "Reagan's Corner." It was there that I was sworn back into the military.

Being sworn in is a necessary formality, when you enter the Armed Forces. Since I had left the Army a few months early, my transition to the National Guard required me to be "sworn in."

Soldiers are usually sworn into the military by the commanding officer or their recruiter when they join the military. Dawn and I had gotten close with our next-door neighbors in Alexandria and one of them was a major in the U.S. Army Reserves. I asked him to swear me in at "Reagan's Corner" at Pat Troy's. He did so and it was a great honor. I was also parting with the conventional means of joining (or re-joining the military). This was doing it my way.

By now, my outlook on life had changed significantly. My primary focus went from returning to Hollywood to getting ready for marriage. Everything outside of life with Dawn was secondary. Building our life together was now the priority.

Chapter 7: Married With Child

"Wise men say / "Only fools rush in," / "But I can't help / Falling in love with you." – Elvis Presley

July 2007 – April 2008

Married

July 2007

After two years of being engaged, Dawn and I finally got married. On July 28th of 2007, we were married by the Mayor in the Woodland Chapel, at the Stroudsmoor Country Inn in Stroudsburg, Pennsylvania, nestled in the Poconos. There was a canopy of trees and a small stream.   Dawn and I walked up the hill and exchanged our vows.

Since he is a lawyer, my brother Matt assisted with our vows. My brother Patrick blessed Dawn and me. My brother Chris and Dawn's sister Paula toasted us. Dawn and I had our military friends provide a saber arch. The saber arch is a military tradition. The saber arch's origins stem from something in the Royal Navy. All that being said, it's a long-standing tradition that an arch is created by the other Soldiers in the wedding by raising their sabers and forming an arch for the bride and the groom to pass through.

After that, our reception went into full swing. We ate and had some drinks and then ate and had some more drinks. The first dance Dawn and I had as married couple was to Jerry Butler and Betty Everett's rendition of "Let It Be Me." We cut our cake to Joe Strummer and the Mescoleros' "Mondo Bongo." We danced again. "Mondo Bongo" is a long song (almost seven minutes). Dawn and I started to wonder when it would end since all eyes were on us, and our guests wanted cake!

Surrounded by our family and friends and cozy in the pristine Poconos, we had a wonderful wedding day.

The next morning we left to honeymoon in Crete.

In Crete, we stayed in a villa that had views of the Libyan Sea. The ocean was warm and beautiful. I taught Dawn how to snorkel, she drove the stick shift vehicle that we rented all over the hills of Crete, and we explored ruins all over the island. We loved it and each other.

Dawn and I were grateful for the fact that some how, some way, in the chaos of war, we met.

Our Bundle of Joy

April 2008

Shortly after we returned from Greece, I was scheduled to attend the first phase of the Infantry Officer Advanced Course. This was due to my National Guard position.

One day, I was getting ready to travel to Fort Benning, Georgia, for more training for the position, when Dawn told me that she thought she was pregnant.

Shortly after I arrived in Georgia, I got a call from Dawn. The doctors confirmed it. A baby was on the way!

Fall became winter, and winter became spring. In the spring, our daughter Meghan was born. After exactly nine months of being a husband, I became a father, and it was incredible.

Dawn's sister Paula, her husband Scott along with my mom and my brother Chris were all present at the hospital. Apparently, after the actual birth, I was so excited that when Scott asked me how Dawn was doing, I simply said, "Everything is okay. I need to get ice."

Of course, Scott, Paula, my mom, and Chris were incredibly confused. Shortly after our conversation, one of the nurses let them know that I was getting ice for Dawn and that the baby was born.

We named our daughter Meghan.   She was a happy and healthy baby.

Following Meghan's birth, I finally felt like I was "home" again.

I realized that everything happens for a reason. Sometimes, we don't always know or understand why we're diverted onto a different path. But with faith and hope, we can persevere through the bad times and enjoy the good times.

The Best Is Yet to Come

I once had dreamed (literally) about being a successful producer. I imagined myself standing in the back of a movie theater, watching the audience watch a film that I produced. I could read the mind of each and every person in the audience. The film connected with them all physically, intellectually, and emotionally. This was a dream that I had for several years.

Going to Airborne School instead of screen testing for "The West Wing" was a tough pill to swallow, but I had to keep my promise. Leaving Hollywood for the Army was even more difficult, but it was my duty. Leading men in battle was a true honor, but combat was good, bad, and ugly, sometimes all at the same time. Nevertheless, all of this led me to Dawn and Meghan. I cannot imagine not having my wife or daughter in my life.

Hollywood taught me that people need people. We are all connected to one another in this life. I learned that we can do anything, but we can't do everything, so focus on the important things.

Combat taught me to take things in stride and to take life, in general, less seriously. Marriage taught me compromise. Being a father taught me that I could have a new dream every day, if I want to. The future is always better and brighter.

Striving to be a good husband and father replaced my pursuit of the dream of being a movie producer.

But more importantly, being a father also brought to life a lesson that I learned from my own father – the most important job that I can ever have is being DAD.

Epilogue: It's A Small World After All!

"It's a world of laughter, a world of tears / It's a world of hope and a world of fears / There's so much that we share / That it's time we're aware / It's a small world after all" – "It's a Small World"- The Sherman Brothers

August 2009

We decided that for our first family vacation that we would go to Disney World. Dawn had been to Disney World once before, but I had never been. Of course, it was Meghan's first time to the Magic Kingdom too.

At the time, Meghan was too young to know all the characters and princesses, but at 15 months, she was amazed at the spectacle of Disney World. It really is a magical place. Everyone is happy – old and young, all races, religions, American and foreign – it is that kind of place. The entire park operates with incredible efficiency and it is immaculately maintained. Another remarkable thing is that Disney World is a tribute to America through and through. The fact that millions of people go there every year from all over the world must not be overlooked.

We went on the "Small World" ride. A beloved, iconic boat ride which also represents a grandiose symbol of unity. Beautifully designed and dressed dolls of children representing different cultures and nationalities of the world. The craftsmanship alone is amazing, but hearing the children of the world sing and watching the puppets dance nearly brought tears to my eyes.

I remembered something that Emmanuel, my best friend from back home who had gone with me on that initial road trip to the Army, once said. "If young children ruled the world, then it would be a much better place."

The song echoed loud and proud through the place and through us. A catchy tune, with a simple message. That we are all one. An anthem to teach our children to help others, acknowledge their blessings, live in harmony with one another to fulfill what the Sherman Brothers had dreamed of when first creating the song, as a "prayer for peace."

My daughter Meghan was amazed by the singing and dancing on the Small World ride. All the children of the world happy and together, dancing and declaring...

"There is just one moon and one golden sun,

Well it's just this world,

And it's for everyone

Though the mountains divide,

And the oceans are wide

It's a Small World after all!"

Acknowledgements

I am grateful to my wife Dawn for constantly pushing me to put ink on the page.

I am grateful to my daughter Meghan for unending inspiration.

My parents provided excellent examples of what it means to be married and to raise children. I am forever grateful for that. My brothers have always been supportive of me and I wish to thank them.

My friend and roommate Dave who introduced me to Ryan (my Mentor); I am thankful for my friendship with both of them. Jerry (The Boss), Harry (the Big Dog), and Marc and Joe, my former business partners, all were critical to my "dive" into Hollywood. I am grateful to have met them and worked with them. Curtis (Woody) my platoon sergeant, was a great teacher and leader, I am glad that he was with the Reapers when I became the platoon leader. Brandon (Gunner) my Bradley gunner covered for me in numerous ways during our time in Baghdad and he was like a brother to me. I am happy that we shared the turret. Any success that I had as a platoon leader I owe in large part to both Curtis and Brandon.

I am grateful to the National Guard Bureau Judge Advocate Generals' Ethics Office for reviewing this project and ensuring that

I stayed "legit." I wish to thank the Army National Guard G2 Security Office for reviewing this book and ensuring that I complied with my non-disclosure requirements for the National Guard and the United States Army. The Army's Office of the Chief of Public Affairs – East also reviewed this book in order to confirm that it was cleared for "public disclosure" and I am grateful for that. This project started in 2011. Since then, I have had several bosses allow me to write this book during my off-duty time; I am grateful to all of them.

I am grateful to LTC Fred Dixon, US Army for conducting a read through and providing feedback on earlier versions of this book.

There is a story within this story as to how my co-author Lila and I met…

My older brother Matt had a good friend from college that lived in Los Angeles for some time. Her name is Camille. When Camille learned that I was looking to move to L.A. and attend law school, she called her friend Julie. Julie had an open room in a three-bedroom apartment in the Westwood section of town. Julie called me and offered the room. This was due largely to the fact that her friend, Camille, referred me. Sight unseen, I had a place. Julie worked in the music industry, but had several friends in show business. One of her friends was a young aspiring writer who was going to college and interning at Fox (the studio). Julie's writer friend was Lila. Lila was to potentially become one of my budding writer clients in 2002 shortly before I left for the army. Thus, years later, she has become my co-author.

I am grateful to Lila for helping me see it through from a series of conversations to a published book.

There are others that wished to remain anonymous that contributed to this project – I thank them all – they know who they are.

Finally, any "crimes" of commission or omission within this book are mine and mine alone.

To be continued,

George Mannix

www.ingramcontent.com/pod-product-compliance
Lightning Source LLC
Chambersburg PA
CBHW011229120626
46549CB00008B/3195